ORDINARY
PEOPLE
DON'T
CARRY
MACHINE
GUNS

ORDINARY PEOPLE DON'T CARRY MACHINE GUNS

Thoughts on War

ARTEM CHAPEYE

Translated by
ZENIA TOMPKINS

SEVEN STORIES PRESS
New York • Oakland

Seven Stories Press
140 Watts Street
New York, NY 10013
www.sevenstories.com

Library of Congress Cataloging-in-Publication Data

Names: Chapeye, Artem, author. | Tompkins, Zenia, translator.
Title: Ordinary people don't carry machine guns : thoughts on war / Artem
    Chapeye ; translated from the Ukrainian by Zenia Tompkins.
Other titles: Zvychaĭni liudy ne nosiat avtomativ. English
Description: New York City : Seven Stories Press, 2025.
Identifiers: LCCN 2024054157 (print) | LCCN 2024054158 (ebook) | ISBN
    9781644214596 (trade paperback) | ISBN 9781644214602 (ebook)
Subjects: LCSH: Russian Invasion of Ukraine, 2022--Personal narratives,
    Ukrainian. | Chapeye, Artem. | Soldiers--Ukraine--Biography. |
    Pacifists--Ukraine--Biography.
Classification: LCC DK5531 .C5313 2025 (print) | LCC DK5531 (ebook) | DDC
    947.706/2092--dc23/eng/20250121
LC record available at https://lccn.loc.gov/2024054157
LC ebook record available at https://lccn.loc.gov/2024054158

College professors and high school and middle school teachers may order free examination copies of Seven Stories Press titles. Visit https://www.sevenstories.com/pg/resources-academics or email academic@sevenstories.com.

Printed in the United States of America

9 8 7 6 5 4 3 2 1

Contents

In memory of Evheny Osievsky,
a talented anthropologist who died defending humanity
on May 22, 2023, in the battle of Bakhmut

1

When Gloom Encroaches

The French public intellectuals of the mid-twentieth century made a profound impression on me as a teenager. In late-1990s Ukraine, electricity and heat were routinely shut off. Back then, it was due to poverty. In the winter of 2023, by contrast, light, heat, water, and the internet would routinely disappear because of Russia's bombardment of civil infrastructure. As a teenager, in my final year of high school in 1998, I visited the cold public library in my town all winter long, and there, sitting in a coat and hat, I read a two-volume edition of Albert Camus's works—in Russian translation because Ukrainian translations, from what I recall, didn't exist yet. Those two volumes were the only copy of his work in my small town in western Ukraine, and books couldn't be checked out. From Camus, I moved on to Jean-Paul Sartre. If I'm not mistaken, it was in Sartre that I stumbled on the example of an existential choice that would make an impression on me—still almost a child back then—for the remainder of my life.

Picture it: It's the nineteen forties, and France is under Nazi occupation. A question arises: What should I—

meaning me personally—do in this situation? Stay with a mother who needs me—her one and only little boy—so much, or join the Resistance? And the answer is this: I don't know what choice I will make until I make it. Only after I've made my existential choice will I become what I'll be going forward. Existence precedes essence.

Little did I imagine back then, as a teenager, that at the age of forty I would face a similar predicament. An authoritarian empire has invaded my country. What should I—meaning me personally—do in this situation? Join the resistance forces, or run away and stay with my family? At the age of forty, my family isn't an elderly mother, but my own two small children who need me—their one and only father—so much.

My sons were seven and nine years old at the time. Age, incidentally, is one of the things that hasn't corresponded to my preconceptions about war, which I picked up from classic anti-war books. For some reason, the books almost always describe people who are young. Estimates vary widely but after Russia's invasion, when Gloom had begun its encroachment on Ukraine, a million people, ranging from eighteen to sixty, ended up serving in the Ukrainian army. Frequently, you hear complaints not about the lack of sex, but about lower-back pain from having to wear a bulletproof vest, or knee pain caused by lengthy troop relocations on foot or prolonged shifts standing at a watch post. Most of the people in the army have already formed their own families. Often, like me, soldiers have small children. Of the soldiers in my company, many have grandchildren.

¤ ¤ ¤

I remember clearly the principal feeling I had in those first days that Gloom encroached on my country. I felt love—an all-encompassing love—and solidarity with everyone I saw and thought about.

Later this feeling began to break down. In those first weeks, we all seemed to be in the same boat. Yet, naturally, different people began to make different existential choices. Now, I find myself having to force my feelings of love and solidarity. When I'm not making a conscious and willful effort, the instinctive solidarity I feel gets reduced to a smaller group: the others who decided to fight, to join the Resistance. I'll try to elaborate on this later.

But back then, in those early hours, days, and weeks of the invasion, I felt love toward everyone and everything around me. We hugged a lot back then. Men, by way of greeting, began to lean in, chest to chest, for a few seconds, which had always been considered unacceptable in macho Ukraine.

In February 2022, I felt a tenderness toward every single person and believed that everyone likewise felt a tenderness toward me. I felt love toward every stray dog, and these dogs, likewise, or so it seemed to me, behaved surprisingly gently toward me in return. I felt love toward every blade of last year's yellowed and withered grass, toward every brick beneath the cracked, wet-grey plaster of my nine-story apartment building. I felt love and compassion toward the little brown wiener dog dressed in a white onesie tied up outside the supermarket. The dog trembled as she peered through the doorway, where her owner was probably stocking up on bread, groats, sugar, and candles in haste. I felt love toward the impertinent tabby cat perched between the flowerpots on the first-floor windowsill of the apartment of an old lady who had, in the past, reproached me for my behavior—"unbefitting such a respectable person as you are." I felt love toward the quarrelsome old lady herself, who had now quieted down and was wiping tears with her wrist as she gazed up at me and asked, "How is this possible?" I gazed back at her, my eyes open painfully wide, and didn't know what to say. I felt love toward the young, freckled single mother from the fifth floor as she hurriedly carried down a car seat bearing her infant. The small, thin woman strained and leaned sideways from the weight, then gulped down a sob.

¤ ¤ ¤

In the darkness of one February night, my wife and I woke up in our apartment on the island of Rusanivka, situated in the middle of the Dnipro River in Kyiv. Explosions shook the walls. You could feel the vibration deep in your spine. My wife peered out into the darkness and said, "*Tse vono*"—"This is it" or "It's here."

Of course, we had a go-bag that we had prepared in advance, as the government had recommended. To be more precise, it was a tourist-style "go-backpack." But we couldn't bring ourselves to believe it. Until the very last day of peace, we didn't believe that this would actually happen. Come on, it just wasn't possible. After everything, in Europe, in twenty-first-century Europe, one country couldn't just up and invade another one, right? It just couldn't. It can't. The days of that sort of thing are over. It simply can't happen anymore.

On that last night of peace, our children slept in a tent. We have (correction: had and will have again) this tradition as a family. Occasionally, if the children have been well behaved, I set up a tent in the middle of the living room and sleep in it between my two sons. They love this very much. I love it, too. We're supposedly "practicing for future hikes when we're bigger." In reality, of course, what matters is that the children fall asleep in my arms. Later, I join my wife on the queen-sized orthopedic mattress as usual. When you're past forty, it's hard to sleep on a camping mat on the parquet. Your back hurts.

The children had misbehaved the preceding evening. They kept arguing with each other. The younger one refused to brush his teeth. But at the last moment, my wife and I had decided to "grant them amnesty" and set up the tent anyway. To this day, I'm glad about this decision. To this day, it makes me happy that, on the final evening of peace, my two young sons fell asleep in my arms. I assumed that only I would remember this fact, but no. Not long ago, my older son, now a refugee, wrote to me from Germany: "When the war is over, my greatest dream is to sleep with you again in the tent at home because I remember that when the war started, we didn't take it down."

"Kiddos, wake up." I pulled them out of the tent in my arms at five a.m. "There's no school today. We're going to Grandma's."

"Right now?" the older one asked in disbelief, half asleep.

"Yes, my kitty-cats. Grandma's missed you very much. She hasn't seen you in almost two years."

We hadn't taken the kids to visit their grandparents since the start of the Covid pandemic, so as not to risk infecting them, since they're elderly. But the pandemic in Ukraine ended in a single day, on February 24, 2022. In Ukraine, we sometimes simply say "the twenty-fourth" without specifying the month or year—and everyone understands.

There are, after all, two different lives now: life before the twenty-fourth and the one after.

¤ ¤ ¤

In the weeks leading up to Russia's invasion, there were fireworks going off every night in Kyiv for some reason. It was unusual. With every round, I would startle and insist to my wife Oksana that someone was doing this on purpose, to scare us. I had been under fire before. In February 2015, Russia—or, rather, the government of the unrecognized "Donetsk People's Republic" controlled by it (now officially annexed by Russia)—shelled the city of Kramatorsk in eastern Ukraine. That bombardment was carried out with empty rocket shells, for the purpose of intimidation. Nonetheless, even from these empty shells, seventeen people died and sixty were injured. It seems that that massive shelling took place on the eve of some important negotiations. I was in Kramatorsk that day, working as a reporter. At the time of the shelling, I was conducting an interview in an aid center for war refugees, located in a semi-basement. I was struck then by the differences in people's reactions. Those who were under fire for the first time, me included, jumped to our feet and ran over to the windows to see what was going on. By contrast, the displaced people who had already experienced being under fire squatted on the ground, huddling against the walls—as far from the windows as possible, as low as possible. One of the rockets hit a building on Marata Street two doors down from where we were.

For several years after this, in peacetime Kyiv, I would still startle at loud noises, whether it was banging at a construction site, a military-parade rehearsal for Independence Day, or even those same fireworks. One time, my older son, who was five years old at the time, was taking a karate class while I waited for him in the gym lobby. Not far away, something let out a pop. I nearly burst into the gym to grab the child in my arms and cover him with my body.

Legend has it that the shortest and saddest story ever written, attributed to Ernest Hemingway, is six words long: "For sale: baby shoes, never worn."

In 2022, in an aid center in Poland for Ukrainian children who had been rescued from territories under fire after the war began, a six-word announcement hung on a wall: "Toy donations welcome. Balloons strictly prohibited."

¤   ¤   ¤

While I spent the final weeks before the invasion in a fearful panic, Oksana, as it turned out, was preparing.

"Don't talk nonsense. No one's going to set off fireworks every night just to scare you."

"But they didn't used to happen."

"You really think that someone came specifically to frighten people every night?"

"I'm scared."

"If you're scared, then do something. For example, pack a go-bag."

So, I packed one. After that, I suggested to Oksana on more than one occasion that we take the children out to Grandma's in western Ukraine "for a week or two." But no one knew if there would actually be an invasion, let alone when exactly. For the second time in two years, Russia had gathered its army at the Ukrainian border on three sides—north, east, and south. Military intelligence kept naming possible invasion dates. Every time yet another date passed and nothing happened, we'd breathe a sigh of relief. Had we, in fact, taken the children out of the city when I was proposing to, we would have returned right before Russia did invade. On February 22, two days before its full-scale attack on Ukraine, the Russian Federation, after eight years of unofficial rule, officially recognized its puppets, the Donetsk and Luhansk "People's Republics," for the first time. At the time, Oksana commented, "Phew. That's all?" No, that wasn't all.

A lot of people were scared and yet still didn't believe that what was coming could happen. In our neighborhood, there used to live—or maybe still lives (since the start of the war, I've been there only once, for two hours)—a well-off elderly man. I suspect that he was somehow connected to the government and had access to information I didn't. With mild sarcasm, I would call him Mr. Positive behind his back. He was always smiling and joking, and believed in the power of positive thinking. Despite his age, he was in excellent physical shape. He practiced yoga and walked in the park with Nordic walking poles every day.

"At my weekend place, I've got a supply of firewood, gasoline, and a generator. We'll survive at least half a year," he would say with a confident smile, convinced that the universe revolved around him. "But everything will be fine. Trust me. You know how I know? I'm planning on going to the mountains to ski. See? I haven't gone skiing yet. And until I've gone skiing, nothing will happen."

Through the end of winter, every time we ran into each other in the neighborhood, I would ask him, "You haven't been skiing yet, have you?" "No, I haven't." And he'd laugh.

That last week before the invasion, when I ran into Mr. Positive, he gave me a wave with those walking poles of his, said something vaguely cheery, and tried to laugh. But instead of laughing, he started coughing. He started coughing so hard that he bent in half and couldn't speak. There was a mute horror in his eyes. And panic pervaded me as well.

¤ ¤ ¤

When, on the night of the twenty-fourth, Oksana and I were woken by explosions, I checked the news on my phone, and the first thing I did was rush to my computer because I had a side gig working from home as a newsfeed editor for a large website. My first instinct was a professional one: We were late with the news. Only after writing up two or three news stories on autopilot did I begin to think about how absurd this was: *What the hell am I doing? My kids are asleep in a tent. There are explosions outside the windows. Invaders are moving in on Kyiv.*

Oksana, unlike me, had been secretly preparing. She had agreed in advance with a girlfriend—our older son's godmother—that, "should anything happen," she would drive us out of the city in her car. We didn't own one ourselves.

Now that the moment had arrived, we needed to find a taxi to make it to this girlfriend. She was in her car, waiting in a string of vehicles. The traffic jams had already started.

Of course, there were no taxis. The kids had long been sitting dressed, waiting patiently and quietly. I kept raising and raising the price on the mobile app, but not a single driver responded. Then, the moment that I said to Oksana, "Tell your friend to leave without us," the phone beeped. A car would be here in a few minutes.

The taxi driver was sleepy. As we were crossing the bridge from the east bank of the Dnipro River to the west one, he asked, surprised, "Why are there so many cars at five in the morning?"

He had been fast asleep in his car and didn't know anything yet.

The driver was from the Zaporizhzhia region: In Ukraine, as is typical of many European countries, people gravitate to the capital for work. While we were standing in traffic on a bridge in the middle of the Dnipro, the driver got a phone call from his hometown. He answered the call.

"Meaning, bombed? And how are all of you personally?"

I was worried that he would cut our trip short—that he would unload us and drive off to save himself. I had, after all, just abandoned my own news editor job.

A gray dawn was setting in. Then, a gray morning arrived. The traffic was getting worse. Ukrainian drivers tend to not be orderly, but that morning everyone was behaving. No one was trying to pass anyone on the shoulder in their expensive Ferrari or Lexus because they were better than the rest. On that day, we all, it seemed, were the same.

The taxi driver finished the trip. I thank him very much for this.

People were helping one another then. In the first days of the war, people were unbelievably decent.

¤ ¤ ¤

In our friend's car, after we'd spent half the day in traffic jams to go a distance that normally took two hours, mostly listening to the radio in silence, Oksana pointed at the monotonously gray sky and said, "The weather's fitting.

If the sun were shining, it would be completely, I don't know . . ."

The entire world seemed gray. Sounds reached us as if muffled by a blanket. There was no fear at all. Inside us, everything had gone mute. I'm a professional writer, and I kept catching myself dwelling on a dumb concern: *I can't find the words.*

I was afraid not so much that a rocket would hit us, but of what the children might see along the way: corpses.

But we were lucky. Twice, the towns that we passed through were bombed a few hours later, behind our backs.

We were traveling in a convoy of several cars. Aside from my son's godmother, I didn't know anyone. My wife sat in the front seat of her friend's car, holding our dog, an all-white Jack Russell terrier named Bianca, in her lap. Bianca couldn't stop trembling. She panted heavily. For years, she'd had her own canine routine: walk the children to school in the morning, then return along the canal through the park, spending some time barking at the ducks on the Dnipro. Now here we were, driving somewhere, unfamiliar smells all around. Her owner smelled of stress, the whole gang smelled of stress, and, as if all of that wasn't enough, there was an unfamiliar dog in the back seat of the car. A few days before the war, our son's godmother had gotten a beagle puppy.

At one point, while on a small road (we were circumventing the traffic on the main highways), a military convoy drove past us heading toward Kyiv. The soldiers were sitting on top of armored vehicles. My wife and our friend gasped simultaneously. As they would recount much later, both had had the same thought: *How many of these boys will survive till evening?*

At another point, we were again driving down a dirt road to bypass the congestion. A tank pulled out of the bushes to meet us and blocked our path. We didn't know which side the tank belonged to. It should have been Ukrainian, but who knew how deep into Ukraine the Russians had already penetrated? There was a lag in the radio news.

We came to a stop and stood parked for a while, facing the tank. No one moved. Then quietly and slowly, so as not to provoke, the whole column of vehicles we were in turned around and drove in the opposite direction.

¤ ¤ ¤

As we drove, I peered into my own soul. I was embracing my children, who were sitting on either side of me, and thinking. Laboriously and slowly, I spun, and spun, and spun thoughts and memories in my head. It was as if oil had congealed in my brain: There was a high level of friction and viscosity in there.

We were well aware that we had been living on a powder keg—from 2014, when Russia, after a meticulous propaganda preparation, had seized Crimea and provoked a conflict in Donbas. Incidentally, back then I also didn't believe that something like that could happen in twenty-first-century Europe. I remember saying to Crimean friends of mine in those initial hours, "Haha, we're being told here that you've got some sort of armed little green men down there—without any insignia on their uniforms. That's fake news, isn't it?" It wasn't fake news. Since then, we've known that so-called Russophobia isn't such a "phobia" after all because the Russian Empire can pull off new craziness at any given moment.

It's just that we didn't think about it a lot. At least, not every day—only rarely.

Five or six years ago, I got a little drunk on wine with a childhood friend of mine, and half-jokingly we started to plan how and which way to flee in the event that Putin attacked. We reached the agreement that our best chances of getting away would be to the southwest, via paths, mountains, and rivers into Romania. The border is the least protected there. (That's more or less what men who are scared are doing now.) I've always written only prose. The only poem I've ever written as an adult began with the line, "When war comes, I'll be a deserter"—yet now here I was, schlepping down bad roads in someone else's car, forced to evacuate my own children from our home.

So I sat in the back seat, embracing my sons as I spun those increasingly sticky and unwieldy thoughts in my head.

What transpires inside a person who has never thought about picking up a weapon, who for years planned to run away and hide, and now faces a different kind of challenge from anything they could have imagined? Why and how do they change their views? What takes place in their head? I've been reflecting on this for over two years now, my two-plus years in the army. I untwist, uncoil, untangle the skein, but I still don't understand it, though it would seem that peering inside your own head would suffice to know your own thoughts.

Perhaps there were key moments. Or perhaps they merely seem key in retrospect.

¤   ¤   ¤

On the afternoon of the twenty-fourth, our convoy stopped to have a meal in a village I was unfamiliar with, at the parents' house of someone I didn't know.

The father, a thin mustachioed villager who looked to be in his late fifties, had kind gray eyes. He had a gold front tooth, a firm handshake, and rough callouses on his palms. He spoke in Surzhyk, the mixture of "nonliterary" Ukrainian and Russian spoken widely in rural and small-town central and eastern Ukraine. He struck me as a living embodiment of the behind-the-scenes, unflashy Ukraine normally hidden from tourist eyes and that I like to write about in my books—the "Ukrainian Dasein," as I call it, playing on Heidegger, "the Ukraine that's really Ukraine." In the evening of that same day, the first day of Gloom's invasion, this villager was mobilized into the army. All the men of the village who had previously served were lined up in front of the village council building, issued assault rifles, and stationed, for the time being, at checkpoints in the surrounding area. As I write this, this elderly villager is in a military hospital. He was wounded in a battle against the Russians outside Bakhmut in eastern Ukraine.

It's partly because of him—a man I'd never met before—that I too joined the army.

I don't know what this phenomenon is called in other countries, but in Ukraine, for some unknown reason, we call it "Spanish shame." It's the feeling you get when someone else is doing something shame-worthy, but you're the one who feels ashamed.

I saw how this old man's young adult son behaved after learning about his father's mobilization. Do you remember how Winston Smith, the protagonist of Orwell's *1984*, was finally broken? When he was threatened with rats, which he was scared of, Winston began to shout frantically, "'Do it to Julia! Do it to Julia! Not me! Julia!'" Do it, instead, to a person he was close to. "Something was killed in your breast: burnt out, cauterized out [. . .] FOR EVER," Smith later reflected.

This young man had just reached the village where he was planning on hiding out. Meanwhile, we were getting ready to continue our drive west with our children. Having already received a phone call that his father had been taken into the army, the young man began hiding behind the backs of the women and avoiding looking anyone in the eye. When I asked him to give us a ride to the highway since our son's godmother would be staying on in this village and we didn't have a car, he quietly replied that he didn't want to drive through Ukrainian military checkpoints on the surrounding roads. A woman, who wasn't in danger of being mobilized, eventually drove us in her car.

I'm not judging. I just wouldn't want that to be me.

On that first day already, it became clear to me that the

brunt of the burden would, as always, fall on the shoulders of ordinary, underprivileged people. In Ukraine, these are the poorer people who reside where they're officially registered as living, mostly in villages and small towns. For the most part, these are country folk—bus drivers, construction workers, supermarket security guards, and the like—the "*the* Ukraine" that I've written about in my book of that title. By contrast, people like me, like the son of that mobilized villager—urban professionals of the megalopolises—can easily just lie low somewhere other than where we're registered and spend an indefinitely long time "waiting our turn." Because nobody will find us if we don't stand up and present ourselves voluntarily. We can "fight on the information front" or "the economic front" (standard self-justifications these days), where we're "more needed" and where we "provide greater value" (also commonly heard phrases since the start of the war).

Yes, I could indeed continue writing every day—in particular, blog posts and reportage—as I had before the war. But then the common people would, as always, be the ones paying with their lives.

Naturally, I'm just improvising here, rationalizing in retrospect.

At the time, all I felt was a stinging Spanish shame.

Certain comments, hand gestures, or facial expressions made in passing, which you later remember for the rest of your life, don't necessarily register as pivotal at the moment they occur. But that sharp emotion of shame for another human being, when you find yourself wanting not just to look away but to turn your whole body away—I registered it immediately, because it was the first strong emotion that cut through the lackluster, deaf-and-mute world after those first twenty-four hours of numbness, those first twenty-four hours of Gloom's invasion.

Over the course of that evening and the night that followed, sleepless with heavy thoughts, I made the decision to join the army voluntarily and stand side by side with the other recruits who didn't or couldn't avoid enlistment. The following day, I reached my hometown, where I was still registered as a resident, and went to the military enlistment office.

¤   ¤   ¤

Until that exact day, I had always considered myself a pacifist as a matter of principle. Since then, I've taken to referring to such a position as "abstract pacifism." Abstract pacifism is the privilege of those who aren't faced with an existential choice and who can allow themselves to theorize. Being a pacifist in a time of peace is like being a vegan when you don't eat. In practice, you don't write a petition against incoming bombs and cruise missiles. Any nonviolent resistance against Gloom will take place against a backdrop of death and torture, including of civilians.

Almost ten years ago, during the 2014 uprising against Ukraine's own authoritarian government, I witnessed people die. It was precisely then—during a power vacuum, when the army didn't know who was leading it or whose orders to follow—that Putin took Crimea. I sometimes imagine Putin banging his forehead against a table in the Kremlin in 2022, regretting that he hadn't seized all of Ukraine back then, at the moment of Ukraine's weakness. As for me, after witnessing the murders of people in the heart of a European capital—and this seemed as historically impossible in 2014 as a war of aggression in Europe seemed in 2022—I began to lean away from revolution and toward pacifism as a theoretical construct.

For a long time afterward, I considered myself a pacifist. I read and reread Remarque's *All Quiet on the Western Front*, Boell's *The Train Was on Time* and "Stranger, Bear Words to the Spartans We . . . ," Vonnegut's *Slaughterhouse-Five*, Hemingway's *A Farewell to Arms*, and Heller's *Catch-22*. I even translated *Satyagraha in South Africa*, Mahatma Gandhi's book about methods of nonviolent resistance, from English into Ukrainian. I was already in the army when I grew curious and Googled: What was Mahatma Gandhi doing on the eve of and during World War II? It seems he was writing courteous letters to Hitler, full of respectful addresses and appeals to rethink things and not fight. (It's believed that these letters were intercepted by British counterintelligence, so that the Nazis wouldn't be able to utilize excerpts from them for their own propaganda.) But it wasn't letters brimming with respectful pleas that stopped and defeated Hitler. He was stopped by the people that picked up arms against him.

It hurts and saddens me today to learn the positions of certain American and German leftists, for instance—particularly since my wife Oksana and I are also leftists.

For example, as a young man, I was such an admirer of Noam Chomsky that I translated his work into Ukrainian, including his essay "The Responsibility of Intellectuals," and I did so voluntarily and without compensation, simply to spread his ideas. Years later, on the first day of the Russian invasion, I felt compelled to write a public response to his position from my phone. It follows, in a slightly abridged form:

> A short letter to some Western intellectuals. To whom it may concern. Please share. I can't write anything long because we're still on the run, with my kids who are right here next to me. So, in brief: Ukraine was not "dragged into" war, it was attacked. Without a pretext, unlike Hitler's attack on Poland. I know other countries have faced their share of foreign intervention, and right now you're witnessing overt Russian imperialism. I don't want to make any flawed historical comparisons, but empires have lost wars against smaller peoples before, and in the end, the Russian imperialist government must lose. When you're being bombed, when you're thinking of ways to evacuate your kids, you have a different perspective than when you're sitting cozy in an office somewhere in Arizona. Yes, Noam Chomsky, I'm looking at you, among others.
>
> I started as a volunteer translator of "The Responsibility of Intellectuals" into Ukrainian—now I'm aghast at how you mention, in one sentence, the lead-up to this invasion: "What happened in 2014, whatever one thinks of it, amounted to a coup with US support that . . . led

Russia to annex Crimea, mainly to protect its sole warm-water port and naval base," Chomsky said. What if the US occupied the Mexican state of Baja California? Before "overthrowing capitalism," try thinking of ways for us Ukrainians not to be slaughtered, because "any war is bad." I beg you to listen to the local voices here on the ground, not some sages sitting at the center of global power. Please start your analysis with the suffering of millions of people, rather than geopolitical chess moves. Start with the columns of refugees, people with their kids, their elders and their pets. Start with those kids in a cancer hospital in Kyiv who are now in bomb shelters missing their chemotherapy.

My doubt began to spread: If prominent thinkers like Chomsky could be so wildly mistaken in analyzing our situation, shouldn't we be reconsidering their opinions regarding events in other parts of the world—opinions that I had previously paid great attention to? Isn't interpreting reality through predetermined theoretical constructs an "irresponsibility of intellectuals?"

I can't shake off the following impression: Western intellectuals, including Chomsky, are really talking about themselves and their loved ones and their own self-interest. Russia is attacking Ukraine, so if you analyze the war from an American perspective, even if it's a dissident American perspective, you're still only going to express a certain kind of self-interest. Going back to Chomsky's essay "The Responsibility of Intellectuals": Calling on Ukrainians to capitulate to Russia right now would be the same as calling on Vietnam not to resist the US in the 1960s because their war of independence was being fought with weaponry from the USSR and China. Maybe it was true at the time that China and the USSR wanted to weaken the US militarily in Vietnam, but for the Vietnamese it was always a war of independence.

Theorizing about geopolitics is a privileged position afforded only to those living in security. From that privileged perspective, you see, at best, general outlines. When you yourself, however, are fleeing and are scared for your children, you get to see certain details up close that you cannot experience from the safety of a secure home thousands of miles away. Many of my leftist friends voluntarily enlisted in the Ukrainian army. I'm fairly certain that more of them enlisted than that of my centrist friends. Nonetheless, in absolute numbers, the majority of Ukrainians who are serving voluntarily are not from the elite. They are construction workers, combine operators, and other people without clearly formed or expressed political views.

I have felt the same "Spanish shame" when watching rallies with direct appeals not to help Ukraine, like the ones in Germany with protest chants like "This isn't our war." No, it's not your war. Be glad that you were lucky and weren't attacked. But to call on others not to help the victims of a crime is the same thing as to step out of the way during the Nazi extermination of Jews and Communists, Romani and disabled people. You, after all, aren't the one being exterminated or doing the exterminating. But we have seen how this disengagement led to genocide and a world war. So now we presumably know better.

It's even sadder to observe the reaction of certain people in the Global South—friends of mine in India, for example, or acquaintances in Africa. One would think that, as former colonies, they would be the most sympathetic with another former colony. But, unfortunately, that isn't the case. For some in the Global West, this is "not our war," and for some in the Global South, this is "not our empire." Therefore, "Mother Russia" should be supported because she's "against NATO"—but this was happening not in NATO territory, but in a former colony of the Russian Empire. I do understand and share many of the anti-Western feelings of the people in the Global South. It makes sense for these countries to balance between empires for their own benefit. But it does not make sense to support an empire like Russia, if only by sentiment, when it is trying to re-subdue its ex-colony. Sympathy for an ex-colony is the least one would expect from others who have known imperialism.

It's easy to hide behind the abstract idea that "the more weapons there are, the more war there'll be" when you yourself are safe. It's easy to say, "Surrender—because we're getting scared" (this is, in essence, what a considerable portion of these arguments boil down to). "Sacrifice yourselves for the sake of our peace." Even if veiled, that's what these appeals mean. You can only say these things as long as you're not the one being attacked and destroyed.

But what if one of your brothers is sitting with his wife in a corridor in February 2022 on the outskirts of Kyiv, with gunfire all around and Russian armored vehicles already driving by the building, and he is panicking when he realizes that he doesn't even have an axe in his apartment, nothing to defend himself with other than kitchen knives and a meat tenderizer? And what if your other brother, together with his young, beautiful wife and preschool-aged children, is under occupation in March 2022, knowing what happened in Bucha, the rapes and the murders of unarmed civilians? Things look a little different then.

Arguments such as, "You, over there, just surrender, so that we, over here, don't start getting scared" are perceived differently from an underground bomb shelter in a Ukraine partially occupied by Russians than in a safe Berlin town square or an Arizona office with air conditioning and a soft sofa.

And at the same time, all those who are helping, however they can, elicit all the more gratitude in me.

Thank you.

¤ ¤ ¤

Who knows if it's true, but I find it quite possible that Putin expected Ukraine to fall in a matter of days or weeks in the face of an invasion by the much larger Russian military force he had assembled. Many people in the West expected a quick military victory for Russia, too. So did some Ukrainians. I can make this last claim with certainty because I myself was one of them.

In the months leading up to the war, I translated a great deal of Western analysis of our political situation. Only one article, out of all those I was assigned, noted that, in addition to the logistical, geopolitical, and military terms of the "war equation," there remained one huge and unknown variable—Ukrainians' willingness to fight—and that it was precisely on this variable that almost everything depended.

It's worth pointing out that this X factor was an unknown for us as well. I was surprised when it turned out we were willing to fight. Yet in hindsight, I should have known, because this was by no means the first time we were tested in this way. Even in the recent past, in 2004, and again in 2013, when authoritarianism reared its ugly head, we openly opposed it. Both times I remember heading to Kyiv's central square, thinking, *so be it if only a few of us show up and we lose. At least I'll know that I tried.* You show up with this thought running through your head, and meanwhile, the square is so full of people that there's no room for an apple to fall!

How had I forgotten that? The professional analysts at least should've remembered it, no? Like Putin's scouts, for example. But perhaps propaganda's main problem lies in the fact that its creators eventually begin to believe in it themselves.

Because I strive to be as honest as I can at all times, let me share something with you. I joined my military unit with my foreign passport in my pocket and an inkling that maybe, within a matter of weeks, I'd be attempting to flee abroad after all. Like many of us, I was convinced that we Ukrainians, for all our fiery resistance, would be crushed in a matter of days, or, at most, weeks or months. We would know that we had tried, but winning was another story.

I mentally calculated that everyone wouldn't be killed that

quickly, and, therefore, that my personal chances of dying were far from one hundred percent. If we're talking about psychology, not biology, the chances of my psychological survival were significantly higher than had I immediately betrayed myself and allowed something to be "killed in my breast: burnt out, cauterized out [. . .] FOR EVER."

So, I fantasized, *we'll head into the pine forests as partisans with all the Javelins the Americans have shipped us.* That's all people were talking about at the time—"Javelins, Javelins, Javelins." This type of weapon meant partisan warfare. Because the state, according to all expectations, was going to fall. The government would flee into exile, or start to collaborate with the enemy, and we'd be left as partisans "*dans la prison des frontières,*" like in the song of the French Resistance. To keep myself from crying, I would hum this song to myself:

> *When they poured across the border*
> *I was cautioned to surrender*
> *This I could not do*
> *I took my gun and vanished.**

Yes, yes, I reassured myself. *Even if we lose quickly, at least I'll feel that I wasn't a hapless victim of injustice, but an active agent of resistance to the injustice.* If one were to rephrase the main thesis of Camus's *The Myth of Sisyphus* and adapt it to the present situation: Even if there is no chance of winning, you have to fight. This will be your rebellion against

---

* Translated by Leonard Cohen in his version of "The Partisan."

the absurdity of life, and it's precisely this rebellion that gives life meaning. You at least tried, even if you lost in an uneven battle. So it goes.

But things went completely differently. They did so because hundreds of thousands of Ukrainians made a similar existential choice to mine—deciding that it is better to be an agent of resistance and run toward Gloom than to be a victim and run away from Gloom. Ukraine didn't fall because so many of us chose to fight, despite the odds of losing.

Not only did Ukraine not fall, it became completely different. It seems that the existential choice determines not only what becomes of the individual, but also what becomes of communities, small and large, including such imagined ones as a people or a nation. (As a leftist, I tend to reject this last word because it shares a common root with "nationalism," and yet there it is, this community that is our nation and our people, something I cannot deny.)

I already see clearly now how our conception of ourselves as a people has changed and continues to change. I would even say that our myths about ourselves have changed. My children, who today are refugees abroad, continue to follow the Ukrainian school curriculum in their studies because we're planning their return. Or is it only that we want their return, or that we dream of their return? No, we're planning it—and we're acting accordingly.

Speaking of school and one's myths about oneself:

Thirty years ago, as a high school student, I studied the history of the "long-suffering Ukrainian people." In essence, this was the history of the gradual colonization of the Cossack proto-state, the Cossack proto-republic—initially and primarily by the Russian Empire, and later also by this empire's successor, with slightly different borders and rebranded as the USSR. Other empires, such as the Polish–Lithuanian Commonwealth or Austria–Hungary, don't fit into the model of the current situation because, unlike the Russian Empire/USSR, their successors ceased to function as empires. What did this look like from the viewpoint of the people of the Cossack Hetmanate and what does it look like now from the viewpoint of the people of Ukraine?

For brevity's sake, I offer four hundred years of Ukrainian history in just a few paragraphs:

An uprising almost immediately after the start of colonization, against Peter the Great. Defeat. New local and constant but unsuccessful uprisings. Peasant revolts (not without ethnic violence, viewed as "righteous" social violence by ordinary peasants). The tightening of the nuts and bolts by the empire. The enserfing of peasants at a time when the opposite process is taking place in most of the rest of Europe. As in all empires, cultural imperialism through education (it's from here, in particular, that the Russian language becomes "native" for many Ukrainians, like English or French in Africa, and Spanish or Portuguese in Latin America). Finally, for a few years an independent Ukrainian state after World War I (which is, incidentally, formally socialist).

There you go!

But no, again we're conquered; the forces are too unequal. Local uprisings led by *otamans*. Famine. Repressions. World War II and a new chance at self-determination. The rise of an insurgent army on the formerly Polish, now suddenly

"Soviet," territory, populated primarily by Ukrainians. (There's a propaganda war around this period as well, that won't fit into these few paragraphs.) After World War II, another two decades of the USSR hunting partisans. Incidentally, the word *povstanets*, "insurgent" or "rebel" in the Russian language, is borrowed from Ukrainian. Finally, in the 1990s, the formal collapse of the empire!

Since then, only a mere few decades have passed, and the empire is encroaching once more.

So, what exactly is this history? Four hundred years of a "long-suffering people?" That depends on how you look at it. My sons now, in accordance with the new Ukrainian grade school curriculum, however slapdash it might be, are studying the history of the "heroic and indomitable fighting people" of Ukraine.

Obviously, both versions of Ukrainian history—that of long-suffering victims and that of indomitable fighters— are simplifications. Each is, to a large extent, a myth. But I hope that the point of the above illustration is comprehensible: Your actions change you and they change your view of yourself.

¤ ¤ ¤

My older son is nine years old—or he was when I still lived with him at the start of the war.

Some things children understand unexpectedly better than you anticipate they will, while other things they unexpectedly don't understand. On the second morning after Gloom's invasion, my older son and I went out for a walk around the village while we waited to get back on the road.

It was warm for February. We were talking about nature. We walked around the village pond. We photographed the sculpture of Jesus. A few times, I stroked my son's head, mentally saying goodbye.

Suddenly, he stopped and looked at me seriously. "Dad, I don't want you to get taken to war."

Reflexively, I replied, "They won't take me, sweetie, because I'll go on my own."

I squatted down so that our faces were level. I was surprised by the extent to which my son had understood the preceding day's conversations about that villager being mobilized. I tried to explain to him that sometimes in life—rarely, and if you're lucky, never—there are situations when you have to do what you have to do, and not what you want to do. Otherwise, you'll be ashamed, both in front of yourself and in front of your little son.

My nine-year-old child seemed not to understand this. That made me wonder if I understood it completely myself.

If I didn't go off to fight Gloom, if I had decided to stay with my wife and children, my wife and children would have accepted my decision. They would have accepted my reasons. Of course, I have to remain with them—because they're what's most precious in my life. If I ran away from Gloom, they would have accepted this. But I probably wouldn't.

Everyone has their own motivations for enlisting. Some think in terms of "patriotism," others of "good and evil." My wife and I, in our youth, were leftist idealists, even utopians. Okay, I'll speak only for myself because Oksana is more analytical than I am: I'm more of an "emotional leftist" than I am an "intellectual" one. Our youth went by in an array of ethical categories, moral imperatives, and musings about justice. There were also environmentally and socially themed tattoos. Oksana and I named both of our sons in honor of national fighters—one a Ukrainian and the other a Mexican.

Then, having lived half your life, you're suddenly faced with a more real and unambiguous injustice than you could have imagined experiencing in your lifetime. What you're facing is almost metaphysical evil. As a novelist, I've always viewed Hollywood clichés about clear-cut good and evil in a narrative with skepticism, but now I need to make a concession: Maybe "good" really doesn't exist. "Evil," on the other hand, has criteria, as it turns out:

¤ when your county is invaded by another country, regardless of how it may try to justify its actions;

¤ when, in the middle of the night, your peaceful sleep is interrupted by bombs, regardless of the geopolitical interests that may be used to try to explain this; and

¤ when your children can be hit by missiles: your own non-geopolitical, little, fragile, dear children.

Fucking masters of war.

Meanwhile, my son is asking, with pain in his eyes, "Why does Dad have to leave us?" "Because, my little sunshine, if I run away now, later I won't be able to look you in the eye, or myself, in the mirror when I shave."

At the same time, I'm tormented by guilt, to the point of pain in my stomach.

¤   ¤   ¤

I don't know whether it was an accident or natural and fitting that I told my son about my intention to join the army first, and not my wife. I remember clearly that I was less scared of going to the military recruitment office to enlist in the army than I was of talking about it with Oksana. I was scared of her reaction, and of my reaction to her reaction.

The last film—or as military guys say, superstitiously, the "latest" film, because there will be more films after victory—that I saw before the twenty-fourth was Jean-Luc Godard's *Contempt*, about the deterioration of a couple's relationship because the wife feels contempt for her husband's cowardice. If Oksana hadn't supported my decision, I most likely wouldn't have joined the army. But I was afraid that if I didn't enlist, not doing so could, over time, ruin our fifteen-year relationship.

My wife and I are not only leftists, we're also feminists. If my decision to enlist didn't interfere with my leftist views—conversely, the decision flowed directly from them—then reconciling what I was about to do with our feminism turned out to be markedly more complicated because the end result of my decision would inevitably be that predictable and patriarchal outcome: "A woman stays behind to look after the children."

Before the war, I was probably one of the best-known male feminists in the country. Ukraine is a rather patriarchal, macho, and homophobic society. When my wife and I decided to divide parental leave between us to care for our newborns, I published a slim book about the experience that became a sensation. Who knew? It turns out that a father can, in fact, take care of his own children, and a woman can, in the meantime, earn money. "Are you not a man?" I'd get asked. "You've become like a woman," one of my neighbors told me. A man and woman changing roles—come on, that's too much!

There's a Ukrainian joke that used to be funny but became sad on the twenty-fourth. A villager is asked, "Why does your wife toil in the garden, and wash the dishes, and take care of the children, and look after the livestock, while you sit on a bench in front of the house in the meantime and smoke?" The villager exhales cigarette smoke and, in a phlegmatic voice, answers, "What if there's a war, and here I am tired?"

But now, a war really had arrived.

In a single day, the chthonic power of Gloom once again imposed a traditional division on the roles of man and woman: The man goes off to war, and the woman takes care of the children.

And she becomes a refugee.

She's forced into single motherhood.

I still struggle with this. As with abstract pacifism, this is what happened: You spend decades constructing theoretical views and wrapping yourself in them, and then the practicality of history arrives and blows it all away with one squally gust.

I still don't know what to think about it all. I'll theorize later, when I once more have the privilege to.

For now, I simply feel wild and intense awe for all women in military uniform without exception. I find all of them attractive. Granted, most of them don't have small children, but some do.

¤ ¤ ¤

I studied philosophy in college and consider myself, at heart, a philosopher. That's how strong that early influence of Camus and Sartre was on me. In Ukraine, in the eighteenth century, there was a popular philosopher named Hryhorii Skovoroda who more resembled the Eastern type of sage, a *sadhu* or a Taoist, than a Western-style one. Or, let's say, the Ancient Greek Diogenes. Like them, Skovoroda was remarkable not only for his philosophy, but also in his way of life. In his old age, he abandoned his career and began to walk from village to village, talking to people. His most famous saying is, "The world tried to catch me, but couldn't"—because he disavowed success and money. Meanwhile, today, his portrait appears on a large-denomination Ukrainian banknote, which I find ironic.

Under the influence of Skovoroda, I coined and tried to popularize the verb and action of *skovoroduvaty*, literally "to skovoroda"—to go walking along random routes, typically through villages and preferably at a distance from well-known locales, conversing with the people you encounter, but listening more than talking. Incidentally, 2022 marked the three-hundredth anniversary of Skovoroda's birth, and I had been planning "to skovoroda" as much as possible. But then Russia attacked. I had dreamed of finally spending time in 2022 in the village house where this people's philosopher died, but the building was bombed by the Russians. After the news of the bombing, in homage to Skovoroda, I wrote the following on Facebook:

"Here, dear intellectual, you're faced with a choice. Who will you become? A lefty who named both his children in honor of those who fought for the people yet himself, in a pivotal moment, ran away from a fight? Or a people's philosopher who decided to share the fate of the people?"

I wasn't thinking of Ukrainians in that moment, but of ordinary people in general. After all, our younger son is named in honor of the Mexican revolutionary Emiliano Zapata.

However, I had made an important mistake, one typical of a male, to which a female friend rightly drew my attention. She, like hundreds of thousands of other women, had become a refugee along with her children. This friend wrote, "Not only men ask themselves such questions. Women do too. But a woman isn't even close to having such a choice. Here are your children, and now you're neither a lefty nor a philosopher, you're just female. It doesn't matter what kind of views you have. I find this very depressing."

¤ ¤ ¤

Oksana is a sociologist. Right now, while abroad, she is researching the practices of adapting and building networks of mutual support among women whom the war forced *en masse*, like her, to become refugees. Suddenly, by necessity, they are single mothers—for some only temporarily, but many are new widows, forever.

In my free time, I help my wife by transcribing audio interviews for her research. Recalling some of these conversations, I feel self-conscious that I'm narrating the story of our family, of our escape, in such detail. Many of her interviewees have more dramatic stories. One woman had Russians shooting at her car while she was in it with her child. Another spent twelve hours in an overcrowded vestibule of a railcar with an infant in her arms. She had to repeatedly ask to change the child's diaper in the train driver's compartment; there was nowhere to put the child down in the railcar itself.

One obviously educated interviewee spoke at length, in detail, and rationally for over an hour about children, their safety, and their prospects. At the end of the interview—as sometimes happens in therapy sessions, when it seems that everything has been discussed and goodbyes are being said—the woman suddenly started chattering emotionally: "I have this cognitive gap. I wasn't planning on being abroad. But I understand that I'm here, and that I'm here for a long time. Just because I have children. My maternal identity ties me to a place I wouldn't choose to be under different circumstances. Professionally, you can be this cool specialist with this unique career, but since you have children, you do what you have to do, and that's it. Essentially, you have no choice. I suspect that you understand me perfectly well." My wife did understand.

Perhaps the existentialists' thesis that "existence precedes essence" deserves further clarification. Even before the moment of the existential choice, you are, to a large extent, already determined, as much by your previous decisions as by your own identity, constructed from both within and without. You're also determined by circumstances—circumstances that didn't depend on you. For example, by such random things as gender, the era in which you were born, and your place of birth. Or by something as random as your citizenship, even if you believe, as I do, that people don't belong to states. Or by the randomness of where you were at the moment when you found yourself faced with an existential choice. I have often thought about the fact that had we personally not been woken by bombs, had we not had to flee our own home with our children, feeling like victims, I might well have made a different choice.

A number of the soldiers in my company were working as labor migrants in Europe and returned to Ukraine after the twenty-fourth. They did so specifically to enlist in the army voluntarily. I wouldn't have been able to do that. There were also a number of noncitizens, who weren't obligated to serve—I'm thinking about Georgians and Chechens first and foremost right now, whose people also suffered Russian imperialism—yet who joined the Ukrainian army.

I admire them, but I wouldn't have been able to do that. What's more, I wouldn't have been able to travel to a foreign country from my own peaceful and safe country to fight against that which I consider evil, just as Orwell once traveled to Spain. Perhaps my choice would've likewise been different had I learned about the invasion from the news, living in some small town in western Ukraine, without personally hearing the bombs, without personally having to flee my home.

A good friend of mine who lives in Paris wrote to me recently: "I saw a guy that looked a lot like you in a café. He was drinking wine and laughing. I was overtaken by a feeling of injustice." Well, then—I'm buoyed by the feeling that right now, we Ukrainians are protecting not only ourselves, but also, possibly, the future of this man and his children in Paris. We're protecting them from a darker world, in which, once again, certain countries believe that they can occupy others, and in which millions of people in various countries on various continents will end up having to face a choice: to flee from home, to hide from mobilization, to leave themselves at the mercy of an occupier who can torture and rape, or to join the Resistance.

¤   ¤   ¤

I was afraid. I was afraid to make a final decision because I would end up having to share that decision with Oksana.

I was afraid of telling her until we had arrived in my small hometown, where I was born and where I'm still officially registered. Only there, while Oksana and I were smoking on the balcony of my parents' house, did I awkwardly say something in the vein of, "Well, you do understand that I'll join, don't you?"

I trembled, waiting for her answer.

But Oksana knew. She probably knew before I did. Much later, when we next saw each other, she described that moment to our friends: "My brain immediately knew what was coming. I was just refusing to accept it." Most of my friends weren't surprised either. Maybe people close to you sometimes know you better than you know yourself.

Oksana accompanied me to the military recruitment office. We walked there holding hands. The weather was spring-like, despite it being February. The frosts would come later. In the military recruitment office, we encountered a humongous line of volunteers.

To reassure our children, my wife said, "Don't worry, Dad will just get sent, you know, to guard . . . to guard roadblocks." I don't know how or why, but as of right now, she basically guessed right (though there are aspects of my position that I prefer to not talk about until after the war).

I've been lucky so far. It's the third year already, and I'm still alive.

But most of the people who joined the army from that same little town ended up in an assault brigade. Many of them are gone already. Thanks to them—thanks to all of those who didn't run away but walked into it—we've held on this long in the face of Gloom.

2

We Must Cultivate Our Garden

Moments of personal weakness began immediately for me and continue to this day.

On my first day in the army, in a basement filled with dozens of folding cots, in a distribution center that had been converted into a large barracks, I burst into tears. That day, another soldier, also newly mobilized, helped me. He was a young guy: eighteen or nineteen years old, swarthy, curly-haired, and handsome. He walked up and handed me a plastic bubble-wrap blanket. His gesture made such a profound impression on me—that a person in this type of situation could muster the strength to care for a neighbor. This young man turned out to be a student in a Greek Catholic theological seminary. One question nagged at him the most. Timidly, he looked at me, swallowed, and asked, "What do you think? If I end up having to kill in the war, will they still be able to ordain me as a priest afterward?"

I, an atheist, had the urge to assure him that not only could he be ordained as a priest, he deserved to become the pope. His actions toward me were those of a novice saint. I wiped away my tears. I was sad to be separated from my family, but my day had become easier and, despite the semi-darkness of the basement, a little brighter.

Since I didn't personally wind up in constant danger, the separation from my wife and children has remained the most difficult thing for me throughout my enlistment—particularly being away from my children. As the stay-at-home parent, I had carried both of them from infancy in a carrier sling on my chest, day in, day out. Until the twenty-fourth, I could and did hug both of them, again and again, every day.

The first two or three weeks after I joined the army, they at least remained in Ukraine. But Putin kept threatening to use nuclear weapons. No one knew which city he would target first. Every night, I had nightmares. Every night, in various censored forms, children died in those dreams. Sometimes it was my younger brother as a child. Sometimes a kitten would fall out a window. Sometimes a military truck would run over our little white doggie. So long as it wasn't one of my sons.

The dreams of horrors stopped on the day that my wife and children went abroad.

But in lieu of nightmares, I was now plagued by good dreams.

In contrast to the terror from which you try to extricate yourself in your sleep with all your might, now you find yourself dreaming that you're hugging your children, and everything is so nice, so bright and sunny, that you begin to suspect that somehow this is too good to be true. "If only this wasn't a dream," you dream yourself saying to your children, but at this thought, your children begin to dissolve. You rush to pick them up in your arms, to hold them close so you can feel them, but your children melt away, and you wake up in tears. Your children are gone, and you are in a fucking military uniform, with a fucking assault rifle under you, sleeping in some fucking train car.

Now the horror becomes your true reality. You turn away from the other soldiers and try to gulp down sobs so that no one can hear.

The most difficult time was always first thing in the morning. You wake up from a nice dream, and then reality sinks in: There's still a war raging.

Then you go out for a smoke. Vulgarity, black humor, harsh jokes about the Russians—they ease things a little. Standing around smoking and bantering, the sense of unity and involvement begins to lift your spirits a bit. You're a small part of something greater, and you're clearly on the good side.

You feel that greatness, as well as the desire at every single moment for all this "greatness" to come to an end right now.

I found myself often telling people that I loved them, regardless of their gender.

Human connection turned out to be the most important part of getting through each day. How grateful I am that we at least live in the age of video calls. At all the bus and train stations, in all the corridors of the barracks, and in all the basement shelters, three-outlet extenders were plugged into the electric sockets, and more three-outlet extenders were plugged into those, and on each one of these a phone was charging. Losing access to your phone was more frightening than losing your gun or ID documents.

On my older son's birthday—he was turning ten, and he was a thousand kilometers away from me, and I didn't know when I'd see him again or if I'd see him at all—I drew a little cake with the number ten on it on an index card-sized sheet of paper, wrote "I love you very much" under the cake, and asked someone to take a picture of me with the drawing.

"Come on, Artem, at least smile," the soldier who was taking the picture said.

But I couldn't. I felt that if I tried to smile, I'd become like the character in the Hide the Pain Harold memes.

That entire day, I walked around with my teeth clenched—so that no one would notice the state I was in. But it didn't work.

"Artem, are you going to lunch?" the soldier on duty asked me.

I shook my head and tried to walk away quickly, muttering something along the lines of, "I'm not hungry."

I went out into the courtyard to be surrounded by the spring trees and flowers. The soldier on duty followed me hesitantly, a Kalashnikov slung over his shoulder. "Did something happen? Did someone die?"

My life has died, I thought. Then, I broke down into full-on bawling.

And he, a tough-looking guy with an assault rifle, hugged me. "You do understand, don't you, that most of us are in the same situation?"

¤ ¤ ¤

When I was leaving to join the army, my children gave me a Rubik's Connector Snake as a gift, the same kind that they had. The plan was that we would make all sorts of creative figures out of the snake and email photographs of them to each other. Naturally, the string holding the pieces together snapped almost immediately. But, for me, these colored plastic pieces, these fragments, became a metaphor for our broken life. On some level, I knew that these Rubik's fragments were the most valuable thing I had now.

Oksana kept saying that, for the children's sake, she didn't want to feel anything. But I, conversely, wanted to feel everything, and to do so as fully as possible. It made me so happy that Oksana, the children, and my parents simply existed. I felt such an acute love, the likes of which I'd never felt before.

I carried around those plastic fragments in my breast pocket, next to my heart, for a long time. Then I lost them for a long time when our unit was urgently relocating for the umpteenth time. I cried over the loss of those little plastic triangles.

Only a year later did I find them in one of the compartments of my backpack. I was overjoyed. Now, once again, I carry them around with me everywhere.

Over the first few months of the war, I'm sure I cried more than I had over my entire adult life until that point. It is not that I'm particularly sensitive. Despite the culture of Ukrainian machismo, a man crying wasn't viewed as a particularly shameful thing anymore.

Our master sergeant is a buff, forty-year-old former judge with the *nom de guerre* Judge Dredd because of his excessive correctness and a haircut like Schwarzenegger's in the first *Terminator*. He described to me how, when the Russians fired rockets at his hometown, he "sobbed like a baby."

Another soldier, a professional martial artist in civilian life, told me something similar. He's a two-time refugee. In 2014, he had to flee his native Luhansk to the town of Irpin in the Kyiv suburbs. In 2022, the Russians showed up in Irpin, and again he had to flee. He cried. He joined the army. His family became refugees. He can't comprehend that his daughter, two at the start of the war and now already four, is growing up, developing, and learning to talk without him. "I'm missing the most interesting part," he told me.

I saw a body-builder-turned-officer suddenly grab the bridge of his nose with his fingers to push back the tears. "On the day that the war started, my wife and I were supposed to make *varennyky*," he explained. The memory of making potato dumplings had made him cry.

In Julian Barnes's *Levels of Life*, there's a story about how a grief that you've already supposedly come to terms with wallops you again out of the blue from where you least expect it. One day, I went to get a Covid vaccine booster shot. As I was filling out a questionnaire at the health clinic, in response to the simple question about my place of residence, I once again started to shed tears.

A few days before the war, we had ordered orthopedic shoes online for our younger son. In the initial days after the Russia invasion, Ukrainian vendors weren't fulfilling orders, naturally. But a month or two later, when my children were already abroad, I got a call from the online store. They asked if I still needed shoes for a little boy. My voice broke when I replied, "No, my little boy is far away now."

¤ ¤ ¤

Occasionally, the pendulum swings in the opposite direction, and, for a moment, I feel euphoric for the fact that I'm no longer simply watching and writing about Ukraine. I can now say truly that I am part of this country that has fascinated me for so long.

The last book that I managed to finish reading before the Russian invasion was by the Tibetan lama Yongey Mingyur Rinpoche. It described how he had been the abbot of a Buddhist monastery in India and a renowned lecturer on meditation in the West, but ran away from his own monastery to become an anonymous wandering monk—like an Indian *sadhu*, like the eighteenth-century Ukrainian itinerant philosopher Hryhorii Skovoroda. From an early age, this lama had been groomed and educated to become a guru (he was recognized as the reincarnation of a deceased sage). But, at some point, the guru sensed that he was involuntarily drawn toward the approval of others, "like a flower turning toward the sun." That was why he ran away from all of it and, moneyless, began to wander, "to skovoroda," namelessly relying on the kindness of others.

In addition to the book's content, I was struck by its structure. Virtually in its entirety, *In Love with the World* is about the reasons behind Mingyur Rinpoche's decision, about his self-discovery, and the first weeks of his life as a wandering monk until his near death from his exposure to the elements. Only at the end of the book, in a few short pages, does Rinpoche mention in passing that he subsequently spent a few years migrating to the north in the summer and to the south in the winter.

I sense that something similar is happening to me right now. When you start asking most Ukrainians about the current war, they too will remember most clearly their twenty-fourth and the preceding and following few weeks. The rest has become a new, albeit dangerous, routine.

The most dramatic aspect of all of this is the fracturing of life itself. A new reality arrives, to which you have to adapt—in order to survive physically and, no less importantly, psychologically.

¤   ¤   ¤

During my night shifts, possibly from the lack of sleep, it seemed to me that whether or not the world would grow darker in the future depended on us Ukrainians. That we, without wishing for this, had ended up in a time and place where we were influencing the future of the planet. That some sort of role as Guardians of the Galaxy had been forced on us. We didn't want others to suffer or face these

risks in our stead because that would threaten everything—
and, possibly, life on earth itself. (Putin was publicly
waving his nuclear dick around again at the time.) We
couldn't place the bodies of others before our own. At the
same time, we couldn't give up. We were forced to exhaust
Putin's dictatorship with our own suffering. If we didn't,
life on this planet would worsen. There would be less
freedom. If Putin were successful, the dreams of other dic-
tators would intensify. Of course, once everything ended,
the "countries of the center" would once more forget about
us, the nation-states of their "periphery." In their histories
of the world, they would once again see themselves from
their center outwards and once again the important things
would be the speeches of the president of the United States,
a wedding of the British royal family, and whatnot.

My thoughts played out against a backdrop of the mun-
dane daily life of a soldier. I tried to convince myself that
I was simply covering the backs of the soldiers on the
frontlines, who knew more from one day to the next and
were gaining more experience than I was. I, for the most
part, lived in relative comfort, in a building with windows
stacked with sandbags, with an electric tea kettle for making
coffee during the night shift and a wood-burning potbelly
stove in the winter, wearing a bulletproof vest and helmet
when out on patrol. That is, my life was considerably safer
than the lives of some of the civilians still living nearby.

A few young guys from the anti-aircraft defense passed
through our unit at about this time—kids who had seen
death up close. Their female officer, as one of the boys
described it, "completely fell apart" when their living quar-

ters collapsed from Russian shelling. One of the teenagers' hands shook as he told this story. But, fear aside, he bravely accepted that he was on his way to a new combat unit.

Our conversations here during the day didn't get into anything weighty. There were no Guardians of the Galaxy or even patriotic propaganda. There was cursing at the Russian fascists. There was joy when the kitchen brought better-tasting food on a given day. In their profile pictures on social media, some of the older soldiers showed themselves with their grandchildren, not with their Kalashnikov assault rifles. Our thoughts and conversations revolved around children and loans, wives and family problems. We were doing what was necessary in the moment. It wasn't the opinions of the "center of the world" that mattered, but our immediate environment. We chuckled over those to whom "nothing has happened yet, but they've already shit themselves." One jovial master sergeant with an auburn mustache summed up his take on risk like this: "Not a single hanged man has ever drowned."

But on the night shift, when you're not allowed to sleep, I would ponder the mysterious fact that we had ended up in a time and a place where we couldn't do otherwise from what we were doing. We couldn't ask people in other countries to put their own lives at risk instead of ours, or that they risk World War III for our sakes. Or even that "normal" Russians fight against their own regime—actually fight, as we once had against our own, with hands and weapons, and not with words from abroad, while we scooped up their shit for them.

Because we wouldn't have risked our own lives either if we could have afforded not to.

¤ ¤ ¤

When I arrived in the army and, before being assigned to a permanent unit, was given an unclaimed folding bed in that horrid basement in the middle of a random company of career soldiers who had been serving since before the invasion, the commander walked around looking for volunteers for the Javelins I've already mentioned. I stood up and exclaimed, "Take me!" But I knew then and there that I was deceiving myself. In fact, I more or less understood that as long as I wasn't assigned to a specific unit, I wouldn't be sent anywhere. That's exactly what happened. The commander scowled and, waving his hand, dismissed me: "You're not from our company."

When, two days later, I arrived where I was supposed to be serving for the time being, I exhaled deeply. I had pictured trenches in the sands outside Kyiv, Irpin, Bucha, or Mariupol (that's what happened to some new enlistees). Instead, the place where I found myself was more like a cheap hostel. It was warm on the twenty-fourth, but a week or two later, severe winter frosts resumed. I reassured myself by saying, "Good, good. . . . A few more weeks here, until it warms up a bit out there in the trenches."

Only gradually, after a month or two, did I begin to understand the extent to which I had been lucky in my assignment. I began to think about how good it would be to stay alive as long as possible. So that my kids could grow up a little, and so that I could keep in touch with them at least via the internet for as long as possible. So that they'd be able to remember me better because they would be older already, and on the other hand, so that, "should anything happen," I would already be just a distant memory for them, a blurred image on a Skype screen. Maybe it would be less painful for them that way—should anything happen.

I realize that these are somewhat conflicting thoughts. I was confused. I still am.

It was precisely then, as I was beginning to get used to the idea of remaining in relative safety, that one morning, as we were learning to use machine guns in a training area, the deputy commander of moral and psychological well-being (a holdover position from the Soviet army) decided to give a fiery and inspiring speech—along the lines of, "Look, boys, other units are already destroying the enemy. Meanwhile, we still have a zero tally. Time to fix this, boys!"

I finished firing my training rounds and lay down on my back on the cold and damp ground, feeling nauseous.

In the evening of that same day, I found out that my refugee wife, now in Berlin, had been ripped off of a large sum. The German capital is famous for how difficult it is to find an apartment there, but Oksana had "by some miracle" managed to do this rather quickly. What's more, it was supposedly a nice apartment, and the rent was reasonable. She was so happy! But, as is usually the case when something seems too good to be true, it was. On a website identical in design to the secure Airbnb and using a link similar to the web address of the secure Airbnb, my wife had paid a security deposit and first month's rent in advance. As soon as her payment was accepted, the site immediately disappeared. The link no longer worked. For a few more days,

Oksana refused to believe that this had happened. "If I just got scammed on top of everything, I'll lose it," my wife cried over the phone. She didn't lose it because losing it wasn't an option: She wasn't responsible only for herself. Two children's lives also rested on her shoulders, as well as one canine's.

That night, for the first time in my life, I had an episode of uncontrollable diarrhea while walking. Luckily, I was in the middle of a bunch of bushes. "Think, think, soldier!" I said to myself out loud, commanding myself to find a solution to this shitty situation. I began to laugh hysterically. A river flowed nearby. The water was still cold, a few degrees above freezing. But I undressed and bathed in the ice-cold water and handwashed my clothes without any detergent. Later, back at our quarters, wearing completely wet clothes, I slipped into the laundry room and then the shower, so that no one would have time to smell the stench.

This might not have happened out of fear and stress. There was widespread food poisoning in our company just then. Maybe something we ate wasn't fresh. Or maybe the cauldrons used to distribute food to various military units had been poorly washed in the kitchen.

¤ ¤ ¤

My first impression of the military as a civilian was, to put it in one word, chaos. That's how everything in the army looks when you're looking at it from the inside. We're building a fortification out of sacks on the inside of a building wall, and a few days later we're dismantling it and moving it to the outside of the wall. I'm sitting behind the wheel of a broken-down vehicle being pulled by a rope. Some old guys in some garage repair it for us for free. The car finally starts, and we immediately give it away to someone somewhere. I had been thinking that it was for us.

But out of all this chaos, something is constantly being born. We run around, we bustle, we load and unload the same things three times, and, out of seemingly nowhere, we've relocated our unit to better, more comfortable, and more protected premises. Bustling Brownian motion, conflicting orders, and then, by some miracle, it's unclear how, military hardware materializes. How exactly this magic takes place is impossible to understand, but it's a regular phenomenon.

There's an anecdote about Hutsuls, an ethnic group in the Carpathian Mountains, who decide to renovate a house. The owner asks, "So, where are the drainpipes?" "*To sia vse zrobyt.*" It'll all get done. "So, why are there still no stairs?" "Don't worry, *to sia vse zrobyt.*" It'll all get done. When the renovations are finally complete, the owner remarks, "I don't know who this Tosia is, but he does good work."

I jokingly began to call the army Tosia.

As one villager who had been a career soldier before the invasion explained to me, "It was always like that, even before the war. In the Ukrainian army, there's no such thing as being handed a plough and told, 'Go plough over there.' You're handed a plough, you start ploughing, and then you're told to go in the opposite direction."

Another master sergeant, with the *nom de guerre* Google because he always somehow knows everything, offered me some relevant advice: "When you're told to do something, don't rush too much. Go have a smoke first. If it's clearly not urgent, make yourself some coffee. Depending on the situation, you can even wait till lunch. Have something to eat because it's completely possible that the order will be canceled. Or you'll be told to do the opposite."

And that's how it would happen. How it continues to happen. For example, my first psychological crisis—no, my next psychological crisis; they fluidly flowed one into the next—occurred when we received word that Russian prisoners of war were about to be brought to our unit. Since I type quickly and well, I was asked to type up document forms for these prisoners of war, not a normal duty for my army position. I had already created files on the computer. I was scared. What if our soldiers, contrary to the Geneva Convention, started to beat these prisoners of war? What would I do? Would I have the strength to stand up for lawfulness? Would I remain silent because "we all saw what the Russians did in Bucha?" I shared my fears with a civilian friend of mine, and he replied, "Just make sure you don't start beating them yourself." I felt torn. At the time, one of my brothers was living under occupation. I had doubts as to what I might do. I was tormented and had nightmares about it.

That time, in the end, I never did see any prisoners of war.

Fast-forwarding a little, the reality of war turned out to be not at all as I had imagined. Initially, with an impression of war based on anti-war books, I was scared that people would quickly turn feral and begin to act like wild animals. But it's been two years already, and I haven't seen this at all. In my opinion, soldiers are, conversely, becoming, I don't know, somehow gentler or something.

One time, we were providing an escort for a young man. The guy was turning thirty years old the following day—a significant birthday in any culture. We noticed this in his documents. Without delving into details, he had been accused of a crime. Seeing several people with assault rifles coming for him, he nervously laughed and commented, "I feel like Pablo Escobar." The drive to his location had been long. We didn't know what we would find there, so we brought sleeping bags with us. On our way back, the night was freezing cold, and the car's heater couldn't cope. Our "criminal" was trembling—not only from the cold, but also from fear of the unknown. I was moved when my partner, an armed military guard, took his own sleeping bag—and this was personal bedding, mind you, personal hygiene— and not only handed the sleeping bag to the escorted man to use as a blanket, but let the detainee rest his head on his lap in the narrow back seat of the car, saying, "Better that you get some sleep. We aren't allowed to sleep, but you are."

Then, so that the criminal's feet wouldn't get cold, the armed guard tucked the bedding in on the bottom, like a mother.

Stray dogs always cluster around soldiers. Cats congregate wherever soldiers are stationed. Army guys, torn from wives, children, and sometimes grandchildren, have a lot of unutilized tenderness. Dogs and cats sense that they'll be fed and petted. The boyfriend of a female friend of mine helped cats give birth a few times in dugouts in Donetsk Oblast. She calls him a "prolific feline father."

In the mornings and evenings, a certain hobbling wild mallard duck usually flies over to our premises. He's grown used to the fact that soldiers will, without fail, give him both breakfast and dinner. We're forever joking around about dreaming of shish kebab; meanwhile, we sit around crumbing bread into pieces suitable for a duck. Lately, the bird has grown emboldened and now walks up closer to people than some domestic hens. That said, he still won't take food from our hands.

We have one volunteer who's physically huge and very strong. He returned to Ukraine from Spain after the twenty-fourth. He used to fight for money and claims that in Guy Ritchie's film *Snatch*, these types of fights are depicted truthfully, but too aesthetically. In reality, he says, the kind of illegal fighting he did is dirtier and more brutal than how it's depicted in the film. In his last fight, he says, he sensed that he couldn't beat his opponent who, despite being slow mentally, was significantly bigger. "I'm beating on him and beating on him, and he just takes it. It's like nothing to him. In the end I have to break both of his legs to stop him from chasing me." After that, he stopped fighting.

One day, this fighter and I were walking together through a pine forest. He grabbed me by the shoulder, signaling for me to keep quiet and stop. Then, he pointed at a tree and whispered, "Look, two squirrels."

I Googled later. It turned out the squirrels were some rare breed, from Ukraine's Red List of Threatened Species.

¤ ¤ ¤

My first field assignment turned out to be like an episode from a movie, not a Hollywood one, but a Bollywood one. In works of art, an unlikely coincidence can come across as unconvincing. One day, we received word that a suspicious man was spotted in one of the more remote district centers. It was unclear whether he was a saboteur or a deserter, but he was reportedly behaving oddly. He was somewhere where he wasn't supposed to be, and the explanations he offered were murky.

My officer, that same bodybuilder who was supposed to make *varennyky* with his wife on the twenty-fourth, was glad to be given this assignment. He was tormented by the fact that he would spend the war mired in bureaucratic routine. (In addition to chaos, by the way, I was struck by the gargantuan quantity of bureaucracy in the army, all these "accounting journals of accounting journals." I was glad to be a private and not an officer, to walk around with a weapon instead of folders of documents.)

All of our army vehicles were in use, and the officer enthusiastically decided to drive to where this suspicious man had been spotted in his own car. In preparation, I was issued my Kalashnikov, cartridges, and a pair of handcuffs, which I didn't know how to use. Someone had to demonstrate for me how to fasten and unfasten them.

We drove through beautifully snow-powdered and frosty forests. The silvery pine trees gleamed under the pale March sun. I started to ask the professional officer about his experience of the war and the army. He sat and sat in silence, then frowned and frowned, and finally said, rather aggravated, "I can see that you're itching to talk. In a few minutes, we'll detain this guy, and I'll let you question him. These guys always have some tear-jerker of a story."

We came to a stop next to a tall man in a dark-blue overcoat and a massive winter hat. He had come out into the road himself so that we would pick him up. In my mind, uncertainly, I ran through the army's rules on how and in what kind of situations to exercise force.

When I got out of the car, this supposed "saboteur" rushed at me and exclaimed, "Artem!"

"Serhii?? What are you doing here?"

The officer looked at both of us warily.

¤   ¤   ¤

Serhii and I had met during the summer of 2021, the "latest" year of peace. It was on one of the most remote and wild beaches of Ukraine, next to the Romanian border. This tall and skinny young guy and I had been scrutinizing one another from a distance. At the end of the first day, we greeted each other in passing. At the end of the second, we struck up a conversation. We both saw that we belonged to the same physical type. The Pentecostalists or Adventists who were camped out in tents nearby kept confusing us: "What do you mean? You're sleeping in a tent while your wife and kids are camped out in a minibus?" "No! Look, that guy standing over there is the one sleeping in a tent. I'm sleeping in the minibus with my wife and kids."

Initially, I thought that Serhii was one of those civic-minded people. He spoke an educated and formal Ukrainian, displayed a sophisticated intelligence, and seemed to want to help everyone. "Hey, guys, does anyone have an air compressor?" someone on the beach would ask. "I do, I do!" At one point, he tried to retrieve a children's inflatable swim ring that had been carried out to sea. Another time, he cautioned, "Hey, buddy! Listen, buddy, be careful! I heard someone saw a viper over there . . ."

It turned out that the case of Serhii was more interesting than mere civic awareness. Serhii was a vegan for ethical reasons and, on top of this, a mystic. While in deep meditation, he would have talks with Jesus. He had even received a personal revelation. Serhii the Mystic was a Christian, but his Christianity was a particular and humane form of the religion. For example, he believed Earth to be a "school" for souls. Whoever didn't come to God during their lifetime didn't end up in hell, but instead returned to Earth to try again, and again, and again. In a nutshell, he was a Christian who believed in reincarnation. When I pointed out that his personal revelation was reminiscent of Buddhism and Hinduism, Serhii the Mystic confirmed that he respected these teachings and considered them compatible with his own—that he viewed them as another way of expressing the same thing. However, he nonetheless considered himself to be a Christian, since he had received this "information" from Christ himself. Jesus, incidentally, had studied in India in his time: He told Serhii about this during one of their talks. Serhii the Mystic combined Christianity with yoga. He described how one time, while deep in meditation, he spent time close to Paradise and felt the "trembling of souls." According to Serhii the Mystic, Paradise was a place where you were unconditionally happy. Here on Earth, something was always required for happiness; there in Paradise, happiness simply was.

Serhii and I spent hours philosophizing as we strolled along the shoreline. We came from opposing standpoints on many things. He told me about his theories on "subtle bodies" and "parasites of souls" on Earth, while I told him about the no less heady laws of physics and the dead-star atoms that we're composed of, about the no less amazing self-awareness of fleeting clumps of matter, and about the human condition and our abandonment into being. Despite the opposition inherent in our views, time after time either he or I would say, "I see you and I are converging in our thoughts." Maybe this was because our seemingly opposite worldviews included a shared tolerance of alternative views, and an intolerance toward intolerance and fanaticism. Maybe we were united in our compassion for everything living. Serhii the Mystic was, of course, more consistent in practice, beginning with his veganism and ending with his genuine nonjudgement of "evil people."

Or maybe our mutual understanding arose in part from our belonging to the same psychological type: Let's call it "The Seeker." I recognize that most people probably don't take the fundamental principles of existence to heart as much as he and I seem to.

¤   ¤   ¤

Back then, on my last morning of that trip, on a strip of beach that is mined today, I got up at dawn to pack up our things and then went for a walk by the water. There was someone else there already when I arrived. It turned out to be him, of course. Who else?

Serhii the Mystic was praying. By the time I came over to say goodbye, he was asleep on the beach. Or maybe he was in a state of deep meditation. That day, I jokingly saved his number in my phone as Serhii Jesus because I didn't know his last name. Naturally, I thought that we would never see each other again.

Then, a month later, a thousand kilometers from where we first met, there was an anti-vaccination rally. As a proponent of evidence-based medicine, I attended the rally in the role of a sarcastic observer. Whom did I encounter among the most noticeable and charismatic anti-vaxxers? Who else?

That time, we also rushed to embrace each other.

And now, for the third time, here we were—again in an unlikely situation and, on top of it, in a region unfamiliar to either of us, somewhere in the boondocks, and once again a thousand kilometers from where we had previously crossed paths. As we were driving Serhii back so that he could provide the army with an explanation, our vehicle first broke down and then, as it was being towed with a cable, also ended up in a traffic accident. The vehicle was loaded onto a tow truck, and Serhii Jesus and I, because we were low on space and with complete disregard for traffic laws, ended up sitting up there, in the strapped-down vehicle. We rode, gazing down at the snow-powdered, silvery forests from a height of two meters and talked nonstop. As it turned out, Serhii had, with no success, been looking for an opportunity to join the army in Kyiv, where there were too many volunteers in the early days. That's why he had headed out into the deserted provinces.

"You're a vegan. You're ready to kill?" I asked.

"Yes," he replied.

Serhii believed that what was transpiring was a global battle between good and evil, and he, as a Christian, not simply could, but had to join the forces of Light. In the first days of Gloom's invasion, Serhii the Mystic had formed a "special praying forces unit" with a few friends. They would find out where the most difficult situation was unfolding at a given moment, then would look up this place in Google Maps and meditate on it—all at the same time, in a large group, to help the forces of good.

Serhii, it seems, decided that you could go ask to join the army wherever you liked. The bureaucracy of the army never crossed his mind. I suspect that when he arrived at the military recruitment office, he also started talking about "subtle matters" and the "trembling of souls." When I explained to Serhii why we had come for him, he broke into somewhat hysterical laughter. "So you were going to handcuff me??"

He spent the night with us and wrote an explanation for the army bureaucracy. The proper procedure for enlistment was explained to him. Serhii set out again—this time to the correct place—to join the army. Once Serhii the Mystic had departed, my officer, the bodybuilder, said, "Hmm, yeah. It's obvious off the bat that this friend of yours is, you know, a little out there."

Today, Serhii Jesus is the commander of a detachment of soldiers. Psychologically, he's coping much better than I am. He is a source of support to me. He calls and addresses me as his Brother in Light at precisely those moments when my soul is the darkest.

Serhii is still a vegan, though orchestrating this in the army isn't easy. When we were marveling at our previous three chance—but, in his opinion, predestined—encounters, Serhii Jesus declared that, for symmetry's sake, there should also be a fourth, somewhere on the eastern border of Ukraine. This encounter would serve as a sign that the Forces of Light were winning.

As I write this, we're both in the east already, in Donbas. The Ukrainian army is, albeit slowly and at a huge cost, squeezing the Russian occupiers out of Ukraine.

So, I'm eagerly awaiting my and Serhii's fourth meeting.

¤ ¤ ¤

One other individual has become much dearer to me since I joined the army, although he doesn't suspect it. Yevhen is an acquaintance of mine from a distant social circle, who I once wrote about in my story collection *The Ukraine*. Yevhen always has sad eyes and tells funny, though also biting, jokes. The following episode from his biography will demonstrate what kind of person he is.

Fifteen years ago, we in Ukraine were fighting against our tendency toward domestic authoritarianism—a war that Ukrainian society is constantly reviving, but one that Russian society has lost for now. Yevhen was, at the time, working for a democratic presidential candidate sarcastically known as "The Messiah." Yet when this democratic candidate, on whom so many hopes had been placed, did in fact become the president, Yevhen left his employment right away. He justified his decision with the explanation that "it's one thing to work for the Messiah and another to work for the authorities." In his words, only those who know how to concoct cunning schemes and writhe like a snake eating its own tail, while simultaneously taking kickbacks, stay on in the administration. For example, one official accepted ten acres of land in an elite suburb. Another, according to Yevhen, bought a little house in the same neighborhood as the president. "And it's kind of hard to judge them," Yevhen sighed, "because if you were told, 'Oh, you've got little kids, here you go, here's just a bit of the national wealth for you, you aren't doing this for yourself, but for those you love,' you would take it and before you knew it, your arms would be up to here," Yevhen said, pointing to his shoulder, "in shit. Forever."

When Putin began to threaten Ukraine with an invasion, a mutual friend of mine and Yevhen's asked each of us what we planned to do. I remember answering that I didn't know. But Yevhen, who had already served in Donbas previously and knew the army from within, was unequivocal in his response: "Am I ready to die for this state? No. Am I ready to become a cripple for this state? No."

But when Gloom arrived, Yevhen spent a week thinking about it, then sent his wife and daughter abroad and joined the army. Maybe when faced with making the actual decision, it was no longer about "this state" for him. I don't know.

Initially, I was too embarrassed to keep in touch with him. Now, we're both serving in the same place, but in the beginning he was serving under significantly more dangerous conditions than I was. Guilt also exists in the army because of the unbearable feeling that you aren't doing enough or are serving in better or safer conditions than others.

This multilevel nature of guilt is, for me, one of the most bizarre and unexpected psychological phenomena of war. Russia attacked Ukraine, but Ukrainians are the ones to feel ashamed. Everyone in Ukraine talks about this. If you're a woman and you have children, you're ashamed that you can't help much in Ukraine's defense. If you can help and are volunteering your time and energy, you're nonetheless not in the army. If you're in the army, then you—like me, for example—aren't at the front where the fighting is most severe. If you're on the front line—like Yevhen, for example—you're an officer and not a private: you sleep on a bed, not on the ground; you're in a dugout, not a trench. If you're a private on the ground in a trench, then you're alive, but your friend isn't anymore.

¤ ¤ ¤

A few days ago, my younger son, who's living in Germany for now, came down with a rather serious cold. Fortunately, he's eight years old already and has outgrown his childhood stenosis. Until around age five, the mildest respiratory illness, and sometimes just a worsening of the air quality, would result, in the best-case scenario, in my son needing an inhaler with hormonal drugs and, in the worst case, in us calling an ambulance.

When there was heavy smog in Kyiv one time, my wife and I woke up in the middle of the night from our son gasping for air. Every minute, I kept running out onto the balcony to peer into the darkness, into the reddish fog. Where were they?? Fortunately, the ambulance arrived pretty quickly. After an injection, the swelling in my son's throat subsided. He was able to fall asleep. My wife and I, drenched in sweat from fear and worry, didn't sleep any more that night.

The rest of the time, my son was completely healthy. Naturally, some sort of stenosis in a child, which isn't considered a disability, wouldn't have been a legal basis for me to claim exemption from mobilization. Yet I suspect that, even a few years ago, instead of voluntarily joining the army, I could have become one of those men who are scared to leave their homes to avoid "catching" a summons, as they say these days.

My wife also says that, a few years ago, she wouldn't have ventured to go abroad alone with two children. Having kids in grade school isn't the same as having toddlers.

Because I can imagine full well how I might have behaved differently myself in other circumstances, I've stopped judging the draft dodgers who sit at home, too scared to go outside before the onset of darkness when the military recruiters stop walking around with their summonses. I know a few such guys. One has serious problems with his stomach, but this condition isn't recorded in his military medical file. Another one has tiny children, so his wife wouldn't manage on her own. A third has hemorrhoids and prostatitis, also not an argument for a military medical commission. I'm fairly calm about these kinds of men.

It's true, I struggle more with the men who have bought documents claiming a fake disability or obtained a fake guardianship of a disabled person and now sip beer from bottles with "Heroes don't die" on the labels. (Unfortunately, these are true stories, including the part about utilizing people's suffering in commercial advertising.) I feel an even greater distaste for the "patriotic" men who sit around waving flags and sipping champagne bought for a thousand euros at a posh resort in Courchevel in the Alps, while singing, "Vova, fuck them up, and we're going to help you!" (Again, a true story, backed up by a video.) It's difficult to discern how it is these men are helping from Courchevel.

But that isn't my point. My point is that I am, naturally, not at all happy that, as an over-the-hill pacifist, I've suddenly ended up in the army. Nonetheless, under the circumstances that arose, faced with the choice that I was faced with, I'm at least glad that I became who I now am. Even if I am a little angry, and possibly too sarcastic, and sometimes depressed, at least I've retained the feeling of personal dignity.

¤ ¤ ¤

A few months after the start of the war—my wife and children had left Ukraine long before—I received two days' leave from my unit for the first time and decided to meet up with some of my longtime civilian acquaintances. One of them, in the company of others, claimed to have become "hyper-patriotic" since Russia's attack. Subsequently, he drank two or three beers, lowered his voice, and being a lawyer, proceeded to explain that, "If you sit quietly, you can sit it out till the end." Maybe he meant this as a joke. If he did, I didn't get it.

It consumed me at the time that, for the first time in my life, I hadn't seen my children in so long—and was continuing not to see them, and didn't know when I would see them. I had this fear that I would die without seeing them again. Meanwhile, here my civilian friends were, all sitting with their families. I remember well how, after hearing the part about "sitting it out quietly," I had the urge to return from my leave prematurely—to my military unit, to my people. Naturally, as a polite person who was still relatively measured at the time, I didn't do this and just stayed silent.

Half a year later, toxic feelings resulting from episodes like this became the main reason I decided to begin virtual counseling with a psychotherapist. It wasn't PTSD; it wasn't fear; it wasn't depression as such. It wasn't even hatred toward the aggressor. It was the painful search for fairness within my own society. Maybe I shouldn't be sharing this, but when you're describing the evolution of your psyche, you should at least try to be honest, otherwise nothing will come of it, no?

You notice on Facebook that an acquaintance of yours is looking for an apartment for himself, his wife, and his children in Kyiv, which is fairly safe by now, and, to make things worse, in your favorite neighborhood—in the very neighborhood where you lived until the war, in fact. You catch yourself feeling a sharp pang of dislike, as if this acquaintance is going to be living your life behind your back. You aren't with your children so that he can be with his. Tens of thousands of people who were less fortunate than you have already died to make his pleasant life possible.

Sometimes I catch the gazes of civilian men on me. Some of them have shame visible in their eyes. Some have visible antipathy, which I ascribe to the fact that my presence elicits either shame or fear that they'll be "raked up" up by me into the army.

One time, another soldier and I went to go take a look at a commercial gym because we were going to be stationed nearby for a few weeks and the commander of our company had given us permission to work out in our free time. We were both in uniform, right after being on duty. When we entered the gym to have a look around, it was full of bodybuilders. They started hiding from us in the corners. They probably assumed that we were from the military recruitment office.

All of these things don't hit you on the head right away. Changes in your psyche don't occur instantaneously, as a result of a single event, but gradually, in response to your reaction to dozens of similar trivialities.

With time, an emotional wall grows.

The wall grows, though intellectually you understand that not all Ukrainians took part in uprisings in the past, and that far from everyone joined the Resistance in France during the Nazi occupation, and that not all Vietnamese fought against the US invasion. That is not how these things happen. After the war, when I have the time again, it will be interesting to research how these topics were discussed in France or in Vietnam. Were they even discussed? I've already stumbled on the historian Henry Rousso's term *resistancialisme*. Within a few decades of World War II, no matter who you asked, everyone had close relatives who had fought in the Resistance. I think that the same thing awaits us Ukrainians after this war as well.

You can, of course, simply opt to avoid complicated and unpleasant topics, but doing so will not make them disappear. On one hand, everyone is impressed by the unity of Ukrainians since Russia's attack. On the other hand, there is often an ambiguous attitude, to put it gently, present in the military toward men who are hiding from mobilization. On yet another hand, no one in the military has any particular desire for their relatives to take risks. But everyone is someone's relative, my wife, Captain Obvious, aptly observed. Then again, I do know one woman who divorced her husband because she lost respect for him when the man tried to make arrangements through her influential relatives not to be mobilized.

Things become psychologically easier the closer you are to the front. In the rear, you might see bodybuilders hide from you behind mirrored columns. Here at the front, you also see bodybuilders, but these carry around machine guns. In the rear, men with disability certificates drive through roadblocks in new Land Cruisers on their way to go skiing. At the front, men drive through roadblocks in battered clunkers on their way to combat positions. You catch yourself constantly smiling at the boys in uniform, and that gets transmitted from your face to your brain as well. Those smiles release endorphins.

You catch yourself walking up to an elderly American surgeon in a military hospital, or to a one-eyed British volunteer evacuating women with children, or to a Norwegian journalist, for no other reason than to say, "Hi, I just want to thank you for being here."

¤   ¤   ¤

With each individual person that you speak to, you switch on your aptitude for empathy and begin to understand them.

Even so, you gradually find yourself feeling distant from men you were once close to, but who now live a completely different life. Conversely, you grow closer to once removed acquaintances who also chose to join the defense forces, and to the people you meet here in the army.

Intimacy arises from your shared existence with these men. I don't know to what extent this intimacy will remain after.

I watched as men I knew fled or tried to flee abroad immediately after the invasion. That's why, when the Ukrainian news started reporting that men that had left the country prior to the war, typically as migrant workers, were returning, I didn't believe it. I thought that this was propaganda on the part of our government to keep up battle morale. But a few weeks later, these men who had returned from abroad began to appear by the dozens in our company. I felt admiration for them. As I've already mentioned, I wouldn't have been able to do that. I wouldn't have found the courage to return to danger from safety.

In an attempt to copy my sociologist wife, I try to conduct interviews with soldiers in my own unit. But it isn't working for me. I query the soldiers about how and why they decided to join the military instead of going into hiding, but their answers are always too brief for in-depth sociological interviews.

"I just couldn't be there," Vitalik, with the *nom de guerre* Czech because he returned from Czechia after the twenty-fourth, responds. He consulted with his confessor; Vitalik is an ardent Catholic. In response to his doubts about joining the army, he heard from his confessor what he was hoping to hear: Protecting one's neighbor, to include with a weapon, isn't a sin, but a virtue.

Another quiet and modest man named Mykola, the father of two children, simply replies, "How could I not?" He didn't tell his wife that he had volunteered. He lied, claiming that he had received a summons.

Another man, who's almost sixty and has grandchildren, did, in fact, receive a summons. "I spent a day lying around. I smoked and thought. Then I decided: Everyone has to die at some point anyway. What matters is how you live before you die."

Yet another one, whom I jokingly call the "rural anarchist," declares that he has an overall hatred for any and all governments. Why, then, did he enlist? "Because I don't give a fuck," he replies, shrugging his shoulders and laughing at himself. Then, to top it off, this "anarchist" gets himself transferred to the unit in the most danger—to an infantry unit serving on the front line.

Here's another example, that of my closest friend currently, if I may permit myself the honor of calling him that. Ihor is known in our unit as a "Man with a capital M." He's a strong chess player and plays the *bandura*, a traditional Ukrainian instrument, well. Ihor lived in Finland for fifteen years, but when Russia invaded, he returned to Ukraine and went to a military recruitment center. Ihor explains to me, "Well, I thought to myself, *how am I going to be able to go to the sauna in Finland when there's trouble at home?*"

I was in Finland once in my youth, in September. I spent time in several houses overlooking lakes and fell in love with the brilliantly colored Finnish autumn. The October trees seemed yellower, redder, and more orange than those in Ukraine. Maybe it's because northern countries grow cold more abruptly in the fall. Now, I keep dreaming of traveling to visit Ihor in Finland with my children when the war is over, and of doing so precisely in autumn.

Ihor seems to share my dream.*

---

* Ihor was killed by the Russians while this book was being translated. May he rest in peace.

¤ ¤ ¤

What has turned out to be unexpected for me in the military guys' responses is this: In the army, unlike on social media or on television, there are no loud and catchy proclamations. Soldiers don't talk about their "hyper-patriotism."

One young villager from a western region of Ukraine, which is typically considered nationalistic, once said to me somewhat agitatedly, "You know, I've been getting goosebumps from the Ukrainian national anthem lately. That's so bad!"

"Why is it bad?"

"Because it means that I'm succumbing to the government's provocations."

I don't know if these are actual propaganda "provocations" on the part of the government, but I, as a leftist, sometimes get goosebumps when I hear the national anthem or select folk songs from a century ago. It is not because I've suddenly become more "nationalistic." It's just that the hackneyed metaphors that arise over and over in these songs have taken on a literal meaning. "We'll lay down our souls and bodies for our freedom": This is precisely and literally what men and women in the military are doing right now. "And we'll demonstrate that we're brothers of our Cossack kin": a sappy little line that I grew sick and tired of in school. But now, you recall what an important role the memory of the Zaporizhzhian Cossacks played for the villagers enserfed by the Russian Empire in general and for the serf-turned-national poet Taras Shevchenko in particular. You remember the role that the memory of the fight against colonization played, and the memory of a free, albeit a tad wild, people. Perhaps this period from three hundred years ago still defines the thing that differentiates Ukrainians so starkly from Russians—Ukrainians with their traditional republicanism, even anarchism, and eternal calls to depose rulers, so different from Russians who were ruled by tsars for all those centuries?

After ten years in exile, the Russian writer Fyodor Dostoyevsky made excuses for his revolutionary youth, becoming a tsarist and a Russian chauvinist. *"Zhydyzhky da poliachyshky,"* "the little Jew-lings and the little Pole-lings," he wrote disparagingly in reference to the exploited minorities of the Russian Empire in his diaries, journalistic writings, and artistic works. Today's Russian propaganda, consciously or not, continues to utilize tropes from Dostoyevsky's later works. For example, the little boy who was supposedly crucified by the Ukrainian military in Donbas in 2014 is startlingly similar to the little boy who was supposedly crucified by Turks in Serbia in *The Brothers Karamazov.*

In contrast to the above Russian tsarist, the Ukrainian national poet Taras Shevchenko, also after a decade in exile, was not only unrepentant for his revolutionary aspirations, but grew more furious at tsarism and wrote:

> *The wheat sowed by the tsars*
> *Will be reaped!*
> *And people will thrive.*
> *The still unborn little tsars will die.*
> *And there will be no enemy,*

No opponent on the renewed land,
But there will be a son, and there will be a mother,
And there will be people on the land.

Yet over-generalizations about nationalities are always a little like "pulling an owl onto a globe," as they say in Ukrainian. An owl is more or less spherical, but fitting one onto a globe is a bit of a stretch, so perhaps I'm incorrect here.

I know only that, in Ukraine, those who personally went to defend us from Gloom's invasion talk about "patriotism" in orders of magnitude less than the so-called leaders of public opinion in the rear.

¤　¤　¤

One time, I had to deliver my military documents to an office in Kyiv. The capital, where the Russian attack was repelled, was by now full of people and open-air cafés. It was full of expensive cars, parked Kyiv-style on sidewalks and road verges.

After finishing my official business, I decided to stop by my home, but was unable to. I was walking toward my neighborhood, and it was like the stereotype from American films about a soldier returning from war. I had thought that this didn't happen in reality. The closer I came to my neighborhood, the worse I felt—physically. My chest hurt.

I sat down in a nearby café that I used to frequent before

the war. While I was drinking coffee, a neighbor that I knew well approached: the big neighborhood macho, a martial artist. He hadn't gone to fight. I nodded my head, stood up, and left to avoid talking to him.

When I reached the border of my old district—and this is an island, mind you—I sat down on a curb, cried a little, and headed back. I couldn't bring myself to enter the neighborhood. My children wouldn't be in our apartment either way. So many people who had been protected by others, including by those who had given their lives, would be walking right alongside me.

It was May. The weather was nice. The weather was too nice.

I returned to my unit, to my own kind.

The easiest place to be is a little closer to the front, where you simply don't see this contrast.

I gradually came to feel closer to the people from all walks of life whom I serve with—villagers, construction workers, supermarket security guards—than the hipsters in cafés, the journalists and artists, and the literary crowd that I used to associate with.

One of my current best friends is Maksym, a rescue worker from Makiivka—occupied by Russia today—whom I later visited several times in Mariupol—also occupied by Russia today. He and I met a few years ago at a literary reading of mine in Sievierodonetsk—yes, now occupied by Russia. Maksym was in the audience.

At the time, I had just written a story about this unassuming and not always alluring but authentic country that I ironically called "*The* Ukraine," and, as I was reading it aloud, the piece was still fresh for me, so I got emotional and started to cry. Maksym brought me some water. We still hadn't yet become real friends. That wouldn't happen until after the full-scale invasion.

When the Russians were advancing on Mariupol, Maksym, along with his wife and teenage son, first abandoned their apartment to stay with friends in a safer neighborhood, and then, when that area started getting bombed too, fled to the rescue station where he worked. That's where they were staying when a Russian rocket hit the building.

Maksym, his wife Maya, and their son Ivan were in the same room at the time of the shelling, but each was impacted differently. Maksym got away with a few broken ribs, which, as far as he remembers, he broke against his own thigh when he was pinned down under the rubble. Maya ended up between two slabs of concrete that formed a little teepee, which is why she didn't even get a scratch. They, however, spent a long time trying to pull their son out from under fragments of concrete, and, when they finally raised one of them a bit, the teenager's legs were both bent unnaturally.

Several of Maksym's colleagues who were also at the rescue station that day never did manage to get pulled out.

While Ivan was convalescing in intensive care, Russian occupiers took over the street and started to come by the hospital. They were looking for Ukrainian soldiers and taking them away to torture them. Maksym came to visit his son in uniform because he didn't know about this, as the internet and phone service in the city were both down. Maksym was lucky and didn't cross paths with the occupiers.

The teenager's internal organs began to die. Meanwhile, the hospital was running out of medications.

Maksym and his wife and friends decided to take the risk. In two cars, driving down small country roads, they crossed from the Russian occupied zone back to the Ukrainian side. Because they were transporting a grievously injured teenager, they were allowed through.

After that, Ivan spent a few months in a hospital in the city of Dnipro. The doctor turned out to be a well-known professor, who was in contact with colleagues abroad. Only once the boy had been stabilized were he and his mother transported out of the country by bus to Poland and, from there, by plane to the US. For over a year now, Ivan has been learning to walk again in the US.

Maksym hasn't seen his son and wife since they went abroad. He transferred to the Donetsk Oblast, where he saves other people, refusing rotations to the rear. Every time his tour is close to ending, he submits a request for his deployment to be continued. I sometimes send spicy mustard and hot sauces to him at the front, which he likes. Maksym, in return, sends me trench candles made by his adult daughter. As they burn, the candles smell like pine needles and crackle, simulating a bonfire. Soldiers, too, are in need of small joys.

"Everything's happening as if we were in a horrible fairy-tale," Maksym said to me on the phone recently. "I still can't believe that any of this is real."

¤ ¤ ¤

In the initial months of the invasion, I too was tormented by the feeling that everything that was happening to us was a horrible fairytale. I too couldn't believe that any of this was real. Smartphones, mobile internet, and war? (When this all started, the first thing that soldiers asked for from volunteers donating to the army was power banks.) Zoom classes with displaced children so that they can resume the Ukrainian school curriculum "when it's all over," video session with therapists, and civilians routinely being tortured in the occupied zones? The twenty-first century, globalization, and postmodernity—and then you recall how in the supposedly enlightened nineteenth century, the so-called century of progress, all the way through 1914, many people believed that wars were no longer possible in Europe.

I couldn't shake the feeling of absurdity. During one night shift, another soldier and I watched, enchanted, as the moon rose over the forest. I was conscious that you experience something that beautiful maybe a few times in your entire life. How, in this beautiful world, are there people who begin wars?

In the village where we were stationed, there was a fog one day at sunrise. A duck was wandering around the yard with her little yellow ducklings. Sunrays falling at a slant through the branches of an apple tree in bloom cut through the air. How, in this beautiful world, is it possible that people attack others and torture them during interrogations?

After you've accepted that the singing of birds and the screams of the tortured are simultaneously possible, a question arises for you as a writer: how to describe this? Because you are, after all, in doing so seemingly transforming human suffering into a text—a paltry text. There's something parasitic about this. Before the war, I used to do street photography, but I haven't picked up a camera since the twenty-fourth because the art of documentary photography struck me as predatory even in peacetime, though I do recognize the need for documentation.

Due to my own moral ambiguity, I avoid writing about certain obvious topics, at least for now. Or I write about them only tangentially. You remember the smile of a modest young man, the father of two children, who, without announcing his decision to anyone, quietly transferred from your relatively safe unit to an assault brigade. You and he ran into each other by chance at the train station as he was leaving for the front. He smiled at you in farewell. A week later, he returned in a coffin. Your unit was given leave for the funeral, but you were too ashamed to attend because you're alive. So you remember his bashful smile.

Maybe it's for good reason that Saint-Exupery wrote *The Little Prince* and Camus *The Stranger* during World War II. Maybe writing in which war is not foregrounded is the only acceptable way to write during wartime. I don't know. How do you even begin to write literature at the same time as someone is screaming in a basement torture chamber and someone else with a leg torn off is dying in a trench? You feel shame for writing documentary texts. Even without your paltry writings, everyone knows that war is horrible. Your writing doesn't stop wars.

I try to justify writing for myself with the fact that these texts will be necessary for history, with the fact that I myself was at one time wanting and not finding Algerian testimonies about the war in Algeria or Vietnamese testimonies about the Vietnam War. I'm confident that these testimonies exist, but the majority of the world knows the stories of these wars only from the position of the countries of the center. Already I see readers searching for stories about the suffering of Russian soldiers because they too are suffering, of course. I would be interested to someday learn what transpires in the head of a person who voluntarily joins an army that invades a foreign country and does so apparently feeling that they are right to do so.

I have one more justification for writing right now. By writing from within the times, as opposed to later, you are at least trying—to the best of your abilities and, of course, with your own personal biases, but nonetheless trying—to show the raw reality of war before this reality becomes subservient to school curriculum propaganda and before the historical narrative is established.

All of these, even considered together, are weak justifications. I know. I doubt myself. I seek balance.

¤   ¤   ¤

Perhaps a stronger justification for writing is the need to show—if not here, then sometime in the future—the people who did what they were forced to do. For example, I hope, though not very optimistically, that there will be less social snobbery in my country after the war.

One time, I was derailed for two weeks by a small episode in which no one physically suffered. We were escorting a military column of army recruits on foot. After the first few months, people stopped volunteering for the army, and, to replace the dead, in the second year of the war, the military recruiters brought on the men that were easiest to get their hands on. Naturally, this could be attributed to both the urgent need and considerations of efficiency: You would spend considerable resources on some guy in a convertible, then he would show up with a lawyer, a rigmarole would begin, there would be time wasted, and, most likely, nothing would come of it. But protecting the country was still necessary. The end result was that they took those who were defenseless.

We were walking through a city in the rear. The new recruits, accompanied by us volunteers who had arrived a year ago, were all a head shorter than us. It was awful. Some of the newly recruited soldiers were missing teeth. Others smelled bad. Maybe I shouldn't be mentioning this, but a number of them also weren't sober. These weren't at all the people that you see on motivational videos online. We, the escort of early volunteers, walked alongside them looking maybe not like elite paratroopers as in the movies, but . . . well, definitely not as unfortunate as those we were escorting. I kept looking at the men the state was forcibly recruiting in the second year of the war, and then looking at us early volunteers, and I found myself recalling the Morlocks and the Eloi from H.G. Wells' *The Time Machine*. Voicing such things is inappropriate, but I'm trying here to be honest.

The worst part happened at the gas station. When we stopped to rest, a cavalcade of two dozen bikers on expensive enduro off-roading motorcycles rolled up to the gas station. Well-groomed, tanned, and buff—that's how the bikers looked to me. They were all thirty or younger and were talking excitedly among each other. They were obviously riding for the fun of it, for the adrenaline. Poorer people sometimes avoid crossing paths with the military because they're scared that there'll be someone from the military recruitment office among them. They're scared of a summons. These bikers were hanging out right next to us, without any hesitation. They joked around, then went to buy food. They probably knew that if they did receive summonses, there were acquaintances with government connections that they could rely on. There were lawyers. When push came to shove, there was money. Maybe some of them had already bought the certificates necessary for exemption from military service; a lot of supposedly disabled people had begun to appear lately. Meanwhile, a man without an index finger on his right hand and a leg five centimeters shorter than the other because of an old injury at a sawmill stood next to me in uniform. It seems that these young guys were no longer concerned that the war would gobble them up. They had people they could hide behind to ensure that they could enjoy life to the fullest.

I do, of course, realize how many people who aren't poor—in particular, friends of mine—also joined the army. For the most part, they joined voluntarily, on principle—to not hide and not try to finagle their way out of it. I, of course, also understand perfectly well that everyone wants to live, including those who are at war, including the disheveled, stooped, and not-so-young men we were escorting. Analyzing such phenomena as inequity during wartime will become more possible when everything is over. Attempts will likely be made to try to forget about this in official history. This is how it has probably been in all wars. There were parties and social gatherings happening in England during World War II. As Orwell wrote, "The lady in the Rolls-Royce car is more damaging to morale than a fleet of Göring's bombing planes."

We stood at that wretched gas station with our new recruits. Some of the soldiers sat down on the curb. The young guys fueled up their motorcycles. This is what I felt at that moment, acutely and with my whole heart: I loved my Morlocks. I hated these Eloi.

¤   ¤   ¤

The Ukrainian army seems to have truly become a "people's army" since the Russian invasion. It now represents practically all strata of society and includes variegated personalities. Its members range from homeless people to IT specialists; from a former gangster who was in prison a few times in his youth to the People's Deputy from my district, a businessman and the sponsor of a children's soccer team; from a grandfather of two young children who had a heart attack not long ago and now walks around with a shunt in his heart to an Olympic champion in track and field.

If I try not to dwell on acquaintances who are hiding from summonses and focus on army volunteers I know, then I reminisce about Kolia, a tattoo artist and anarchist. I examine my forearm and am glad that Kolia, together with his brother, joined the Territorial Defense in the first days of the invasion. I'm proud of the fact that a little part of him has remained under the surface of my skin. I think that it would be less pleasant to look at my own arm every day if the person who had adorned it elicited negative emotions in me. Incidentally, three out of the four employees of this left-wing tattoo parlor joined the defense forces, to fight injustice.

Not long ago, I by chance saw Fedia, an acquaintance from when I was young, in a video from the front. He was once a lecturer in philosophy at a university. Then he decided that a true thinker should be among the people. He resigned from his department, studied to become a skilled worker, and went to work on a tower crane. He is the same sort of philosopher-practitioner as Hryhorii Skovoroda, who moved from the realm of royal courtiers to the realm of village itinerants in the eighteenth century. Only Fedia, living in the twenty-first century, is rightfully more urban. Right now, he is a field scout in the city of Bakhmut, infamous for Russians' crimes and virtually destroyed by them. In the video interview he describes the city:

> Bakhmut looks as if it's been stirred up with a spoon. Fragments of houses, burnt equipment, poles, trees. There are probably people somewhere in this mixture too. Craters that vehicles sink into . . . And in the background of all of this, cherry trees are blossoming. You move through Bakhmut in the morning at a run with guys from your unit. Everything's in ruins. It's blan-

keted by either morning fog, or powder gases, or the smoke of fires. Everywhere there's fragments of shells, chunks of equipment, and broken trees. And on the surviving branches, it's as if nothing's happening: Cherry blossoms are blooming. And you have this impression that you've ended up in a stupid dream, where everything is happening not as it should. I'll remember these Bakhmut cherry trees for a long time.

¤   ¤   ¤

Many people who aren't in the army themselves—I'm thinking now, above all, about women, including women with children—also help the army however they can. My wife spends a quarter of the family income every month on donations to various military units. A car recently went up in flames in our company (not from shelling, but due to old age). I wrote a Facebook post about it. Three hours later, enough money had been collected for our company to buy another one. This is a regular occurrence.

The police have also stopped being perceived as separate or alien. I go with policemen on joint assignments sometimes. When a cat with newborn kittens died before our eyes, young policemen nursed her five kittens from a syringe.

The army and the defense force are, first and foremost, the "people's" in the sense that, with the exception of the plutocrats and their children, they seem to have become a more or less exact cross-section of society. For instance, in our company—and, as far as I can see, the situation is similar in other units—there is one intellectual, two professional athletes, one man who was in criminal gangs when he was young, two or three businessmen, and one professional artist, as well as a few amateur ones. But the majority are, of course, manual laborers and peasants.

This is, naturally, an arbitrary categorization since people simultaneously have different roles and identities. A professional mixed martial arts fighter can simultaneously work in crypto mining and securities trading (a true story). Someone who was a pickpocket for the adrenaline rush and an adventurous hitchhiker when he was young can simultaneously have been trained first as a pianist and second as a drilling engineer (also a true story). A peasant can travel abroad for work and invest the money he's earned in speculative construction back home (again, a true story). All of these men are fighting against the Russian invaders right now.

There are also differences depending on where in Ukraine a specific unit was formed. In my unit, where people from western Ukraine prevail, specialists from various fields of construction and renovation seem to dominate. However, in the unit of one of my acquaintances, which was formed in the steppes of the south, combine operators and oil press operators prevail. This acquaintance has an MBA, is philosophically inclined toward the ideas of Ayn Rand, was a Fulbright Scholarship recipient, and studied in California. He jokingly told me, "These people are like children— these combine and oil press operators here. Yes, especially the press operators. . . . They sit down around a bonfire and begin: "Oh, tell us something else about San Francisco.""

"Tell us another story." "So, you've really visited the gay quarter? Wow! So, what did you see there? Real gay guys?"

Since the beginning of the war, intellectuals, businessmen, hipsters, artists, "successful people," and journalists in the army have aroused the greatest interest—among intellectuals, hipsters, artists, "successful people," and journalists. The "out of the ordinary" military guys get interviewed, they write essays, they shoot videos that they post online, they reflect, they describe their experiences, and they represent the reality of war. But all of them—more precisely, all of us—constitute only a tiny minority of soldiers. The majority of Ukraine's military personnel remain unrepresented, invisible, and voiceless.

Yesterday we were stationed at a roadblock. A seasonal village fair had just started nearby. Upon noticing it, some of my colleagues grew happy—because it's spring, and they need seedlings right now. The soldiers won't be able to do the planting this year themselves, but they can at least pass along the young saplings to their families. War or no war, we have to "cultivate our own garden," as Voltaire commanded.

Standing on duty at the side of the road, I watched my companions and loved them. Without any "buts," I loved them. Two middle-aged men with machine guns slung over their backs, with half-smiles on their faces, a little excited and a little languid under the first warm rays of the March sun. The older one was hobbling; because of an old industrial injury, one of his legs is shorter than the other.

The soldiers spent a long time wandering among the people in the bazaar, meticulously inspecting the seedlings.

In the end, they chose a peach tree and a cherry tree.

People Can't Be Separated into Varieties and Breeds

I was scared of how the war would change me.

Stereotypically, I believed that I would become more hardened and crueler. In order to counteract this, I deliberately tried to change in the opposite direction—to become more sensitive, to try and be kinder. I began to buy sausages in village stores for stray dogs and cats. In May, I spent an entire month feeding an anthill in the woods. I would bring pieces of apple and the soft filling from chocolates, and then would watch as the ants enveloped them.

I appeared to be successful at not becoming more hardened, but war did, nonetheless, begin to change me, in unexpected ways. For example, no matter how much my wife criticizes me for this, I haven't been able to prevent a dichotomy from developing in my perception of those who went to fight injustice versus those who didn't.

I used to appreciate that I had retained a group of friends back from high school and university that I was still close with. There were a dozen of us. We were godparents to each other's children. None of us are religious; in Ukraine, serving as a godparent to someone else's child indicates great closeness. No one from this circle of "crisscross god-parents" joined the army. I continue to keep in touch with each of them individually because I can understand where each one of them is coming from. But it's become hard for me to identify with these friends from my youth as a group. Instead, I find myself identifying more with other social groups, like the circle of leftist friends that my wife and I share, most of whom went to fight.

I also still don't know what to make of the division of responsibilities between men and women going forward. Despite the feminism of my pre-war social circle, men went off to fight, and women stayed with the children. If being killed or disabled fighting in a war is the "price of male privilege," it is a rather high price to pay. At the same time, there are many feminists among the women who joined the army.

Maybe someday I'll regret having expressed half-ripe thoughts here, but it seems better to be honest than infallible.

¤   ¤   ¤

I have barely mentioned Russians up until this point.

The majority of Ukrainians don't struggle with their attitude toward the enemy; their feelings are obvious. Those who reflect less hate and dehumanize Russian soldiers. For instance, in the initial months of the invasion, Russians were compared to the orcs from Tolkien's books. This was bad, in my opinion, but I can understand the emotions of the people who spoke like this. Those who are more prone to reflection are more likely to pity Russians without pity. I don't know if a person living in a peaceful country will be able to appreciate the full meaning of this feeling, but one of the strongest artistic expressions on the subject of the Russian invasion, written in the style of a traditional Ukrainian folk song, includes the line, *"Oi zhal meni, vorizhenku, shcho ty stav na tsiu dorizhenku."* "Oh, I pity you, little enemy, that you're standing on this path." The poem was written by Anastasia Shevchenko, a female military volunteer.

Not long ago, I transcribed an interview conducted by sociologists with members of the Russian opposition. Anti-war protests were quickly suppressed in Russia. I was struck by the story of a certain man, who was against the war, felt lonely, and fell into despair out of his sense of helplessness. He set the door of an FSB building on fire and didn't run away. He wanted his action to become an example for others. But the people in the vicinity turned away their gazes. They lowered their heads. They walked by quickly. They pretended not to notice. The man stood there alone until he was detained.

What a sad dystopia.

I wouldn't want to live in that kind of society.

¤ ¤ ¤

My feelings toward Russians have changed over time, from rage to disgust, and then to pity. At first, I naïvely anticipated that Russian society, indignant over the injustice of the invasion, would bring down Putin's regime. Now, I feel pity toward Russians for having to justify their conditional membership in that kind of society.

In the first weeks of the invasion, I too referred to Russian soldiers as orcs, if not publicly. Then I started not to like this. My thinking was as follows: *Orcs have no free will. They're creatures that were created for evil. The majority of Russians are people who reject free will, or who don't have the strength to resist evil. It is the fact that they are people that makes their responsibility even greater.*

I have a friend who offers group therapy sessions. He recently shared something with me as a therapist. You can gradually replace all of the participants in a particular group down to the last one, but the rules that existed in the original group will remain. The only way to change the rules is to break up the group and then assemble it anew. But at the level of a huge country, this is hardly possible.

There is a well-known fable about a frog. Allegedly, if you toss a frog into boiling water, it will instantly jump out, without having time to get scalded. But if you place a frog in cold water and begin to slowly heat it up, it will acclimate to the temperature changes little by little but will eventually get cooked. Perhaps something similar has happened in Russia. And this seems to apply to Ukrainians who have lived in that country for a long time as well. There are close to a million Ukrainian citizens in Russia, and they too were unable to do much. By contrast, there are Russians voluntarily serving in the Ukrainian army.

It is, therefore, not a question of ethnicity or a passport. It is more that Ukraine has, over the course of its history, developed a society that's far from perfect but is nonetheless a little healthier. Ukrainian society has domestic examples of successful protests and imperfect but nevertheless positive changes.

Ukrainians are not elves, and Russians are not orcs. Human beings constitute both sides. That's what makes the situation tragic.

⊠　⊠　⊠

Ukrainians are transforming from a people almost everyone expected to flee and fall in three days into a people coming to terms with a lesser perfectionism in our goals.

Simultaneously, Russians are becoming a community where the majority rejoices at this attempted genocide because genocide and imperialism lend meaning to their existence—and in the meantime, a minority feels a burning shame in response. In July 2023, seventy percent of Russians said that they supported Putin.

The stigma will remain for a long time, perhaps for generations: the need to hide that you're Russian. The Russian rapper Vladi, in a rare example of self-reflection, describes in one of his songs Russians who fled to Thailand, who don't speak Russian out in the street and who try to pass for

"Europeans with no distinguishing features." There's that need to distance oneself from one's people or to offer justifications. There are those "bearers of God" for whom you feel wild shame.

Because you become what you do. And to dilute the unintentional pathos, I will add one thing. I would very much like for Ukrainians to become not the "heroic fighting nation" depicted in our new school curriculum, but more like . . . well, for example, the Swiss—as I imagine them to be, of course.

But history, unfortunately, hasn't afforded us those kinds of circumstances.

¤ ¤ ¤

When I was in preschool, I was forever drawing airplanes in flames, and sometimes tanks also. They were Nazi planes and tanks. I would schematically illustrate this fact with Wehrmacht crosses on the bodies of the burning equipment.

I was born a third of a century after the end of World War II. Not even my grandmothers and grandfathers were among those who fought. But, in a Ukraine fully occupied by Nazis, that great war touched practically every family in one way or another. My paternal grandmother, who was eighteen at the time, managed to escape into a forest when *Ostarbeitern* were being rounded up throughout Central and Eastern Europe to be taken back to Germany for forced labor. Her mother was taken in her place. When the wagons carrying the gathered people were rolling out of her village, my grandmother, then a teenager, came out of the forest to meet them, saying, "Get down, Mother. I'll go." My other grandmother, who lived past the age of eighty, would tell and retell and again retell the exact same story through the end of her life. A Nazi had put a noose on her as a four-year-old child because she had "stolen" (i.e., hidden from the German soldiers) an egg from her chicken coop. Another German, however, saved her from hanging. Grandma talked about that noose around her neck until her death. For over eighty years, she tried to "talk this memory out of herself," so to speak. She was never able to.

As I recall myself in childhood with those drawings, I often think about the fact that perhaps my grandchildren and great-grandchildren will, as children, draw burning Russian planes and tanks the same way for another half a century. This makes me very, very sad. I would like to live in a country where children no longer have a mental image of war and an enemy. But not everyone is so lucky.

¤   ¤   ¤

Certain old friendships of mine are dissolving because the points of tangency in our lives are disappearing, while simultaneously topics that are difficult to talk through are appearing. Maybe this is temporary. I don't know.

Both of my children are losing their baby teeth right now. I see that in the pictures my wife sends me from abroad.

The process of a relationship dissolving unilaterally reminds me of how the roots of baby teeth dissolve. Baby teeth fall out, but other ones grow in their place. Some of the people I have met who have served alongside me in my military unit, as well as some who I used to only know in passing and have gotten to know better during this war, have become the most important people in my life.

Serhii Jesus, whom I had driven to "catch" at the start of the war, called me not long ago. He's one of the dearest people in the world to me now.

Maksym—the rescue worker from Donetsk—and I are constantly exchanging gifts by mail these days.

There are things that one military man will talk about with another military man before he will with anyone else. Because he isn't sure that someone outside the military will understand him. Because the deepest of experiences for you is a subject of passing interest or small talk for a civilian or a foreigner. Someone in the military begins to share what they consider to be innermost, and someone outside the military devalues it with one awkward remark. Or, after two sentences, that person has moved the conversation to something more interesting, like who's currently in the running for an Oscar or a Grammy. I've encountered this. It's painful.

One of the drivers in my unit, who's a construction worker in civilian life, once said, "I went home on leave. People were asking me, 'So, how is it?' But what is there to talk about? He who's been here—he knows regardless. He who hasn't been here—he won't understand either way." That's why military personnel talk about their experiences for the most part with other military personnel. But even here you can make horrible mistakes, as I did recently.

I could see that he was eager to talk to me: a fifty-year-old man with a pointy beard, sitting at a neighboring table in a village café. We were both in uniform; everyone else around us were civilians. I could see by his timid look and his movements that he needed to talk to someone. At first we talked about general things: where each of us served, where each of us had been. Then he said, "The hardest thing is when you end up having to take that first shot at a human being. I saw his eyes . . ."

"Have you had to kill many?" I asked reflexively, then bit my tongue.

"Dumb question."

"Sorry."

"Dumb question. Sorry. Sorry. Sorry . . ." His eyes were filled with tears.

"Sorry."

"Sorry. Sorry . . . Never ask that sort of thing. No one will ever tell you."

He spent a long while recovering from my dumb question.

"I didn't sleep for a month. A month. Lord, forgive me, a sinner. As soon as I fall asleep, I see his eyes. On leave at home, the cat comes into the room and I flinch. People like us . . ." and he again swallowed a sob. "They were thrown at us like dogs. If I'm not the one to . . . other guys will have to. Then you come back, and you're asked, 'Did you have to kill many?'" A sob escaped his lips.

You're sitting across from a person who's killed. Not a "killer"—I'll hit anyone who dares say such a thing. A person that *ended up having to* kill. There he is, half a meter away from you. And you know that he's killed, and you love him more than anything in the world. You love him so much that you feel the rush of hormones—oxytocin, or whatever. You feel the rush of energy that flows from you to him, and you want to hug him, but you're scared. You're scared to hug him. You're scared to touch him because even a cat at night causes him to flinch. You're scared to hug him because the whole thing will turn out more dumb.

¤ ¤ ¤

Sania is one of many. Each has their own story, and each story is unique. But each of these stories is multiplied by thousands, tens of thousands of times. These people will not be interviewed. Even if you were to try, they wouldn't tell you much. Either they'll remain silent because civilians wouldn't understand them, or they won't be able to describe clearly what they've learned and what they've been through. They won't be able to effectively shape their feelings into words. They won't write an essay.

One of them tells his wife a bunch of lies, that he's supposedly been drafted—but he goes of his own accord. Another one finds meaning, perhaps, after years of family problems. Once in the army, one of them is trying to refuse a mandatory vaccination because he believes in the big-pharma conspiracy. Another does not get along with his commander. Yet another might be a run-of-the-mill homophobe. Another may be gay and unsuccessfully trying to ensure that his partner at least has some rights as a partner in the event of his death. One may hate the state as such and be unable to explain what in the world drives him— but he acts. He acts. All these people act.

And here, on the backs of all these non-intellectuals, these non-artists, these completely ordinary but unique personalities that won't express themselves publicly—on their backs rests the future of Ukraine.

And perhaps the future of the world.

¤   ¤   ¤

More often than you would like, you find yourself listening to the opinion that it isn't worth having "special" people— writers, or musicians, or who the fuck knows—play a direct role in the war. Sometimes similar things are also said about journalists, for example, or athletes, or other public figures. A long list of professions follows, from IT down to biology.

I am comfortable accepting people who admit honestly that they're scared of death. I'm scared of it myself. It's less comfortable to accept the explanation that someone considers themselves "more useful" in other roles. Sometimes metaphors like this are used: Utilizing a person of culture (an athlete, an Arctic explorer, insert your own profession) in the army is the same as hammering nails with a microscope. The first obvious flaw with this argument is, of course, subjectivity. Anyone can pronounce themselves a super-sophisticated tool and in this way justify avoiding enlistment.

Some Ukrainians say that they "value their freedom" or "don't want to deal with bureaucracy." Fine. Then add the caveat that this freedom is maintained at the expense of others.

But perhaps the hardest to stomach is when people start using metaphors about "fighting on a different front," like a "cultural front," or an intellectual front, or an informational front, or even an academic front. Where does it end? I saw one advertisement for a Ukrainian bookmaking office (!) that proudly claimed, "We're holding strong the economic front!"

If you've already chosen not to take part in the fight against injustice, why search for additional justifications? Is the fear of death not enough?

Yevheny Osievsky, a good friend of mine, a cheerful and self-deprecating doctoral student in his civilian life, used to write funny anthropological texts before the war. After the Russian invasion, he went to fight and became a paratrooper. In February 2023, Yevheny wrote on his Facebook page:

> I don't doubt that when they say, "People like you don't belong in a war" or "Let others fight," they mean it as a compliment. Maybe as an expression of support. I don't need that kind of support. It takes as a given that writing texts or programs and abusing the word "discourse"—or whatever else it is "people like me" do for a living—is a higher form of activity, a more elite one, than "merely" going off to protect one's family and city, than "just" being left armless up to your elbow, than "simply"

becoming blood-colored foam on the tracks of an enemy tank. People can't be separated into varieties and breeds.

On May 22, 2023, the Russian occupiers killed Yevheny Osievsky outside Bakhmut.

May he rest in peace.

¤ ¤ ¤

Sometimes I'm asked about war fatigue, and I don't know what to say. It feels like the same sort of theoretical construct as the idea of pacifism in the face of bombs and missiles. In the first days, my biggest dream was exactly what it is now: for this nightmare to end as quickly as possible. Exactly as it is after two years of war, my biggest dream on that first day was to take off my uniform and never again in my life put it on. Most importantly, I dreamed that my children wouldn't end up having to don one when they grow up. When one of my sons was still in preschool, he was supposed to play the tin soldier from Andersen's fairy tale in a Christmas pageant. When I saw my child in a toy uniform—which was, nonetheless, a uniform—I almost broke into tears and whispered, "Son, may you never have to wear a real one." Maybe it's precisely because of my aversion that I ended up having to put on a uniform.

I'm in my third year of service and don't know how much longer I'll have to serve. I once Googled how long Americans served on the draft in Vietnam. For the most part, it was exactly one year—three hundred and sixty-five days. I read that some soldiers had a little calendar inside their helmets and would cross off the days. I was jealous. Then it dawned on me that we Ukrainians who are fighting in this war are more like the Vietnamese in the Vietnam War than the Americans. We need to survive. We need to defend the opportunity to make our own choices. Be it the right one or the wrong one, but our own choices regardless.

What does war fatigue mean? If we are fatigued, what of it? What are our options? I don't want to get into an expansive sketch of Ukraine's history to an audience of foreigners, but for Ukrainians, the options are simple. We can either fight back now, with the losses that necessarily accompany this, or remain the colony of an empire for another hundred years. The entire nation won't be able to run away. If we surrender, Russians will be able to do whatever they want with us, as the government of China does with Uyghurs today. Right now, we're receiving assistance in the form of weapons, but if we surrender to the mercy of the occupier, we will give the world permission to merely "show deep concern." Not long ago, the European Parliament finally rejected what I consider to be false political correctness and recognized the Holodomor, the manmade famine of 1932–33 in Ukraine, as a genocide. Of course it's difficult to imagine that someone would be able to kill millions of people in present-day Europe again, but until the twenty-fourth, we couldn't imagine that one country could invade another in present-day Europe either. British intelligence made public Russia's initial plans, which, after the occupation, aimed to divide Ukrainians into four parts and physically destroy those who refused to accept the occupation: liquidate them, just kill them off. We now see how they're deporting children from the presently occupied territories and trying to "reprogram" them. It's for this—not the actual invasion—that an arrest warrant was issued for Putin at the International Criminal Court in The Hague.

I mentioned how unpleasant it is to watch certain rallies abroad call not to help Ukraine. It's easy to hide behind arguments like "the more weapons there are, the more war there'll be" when you're safe. But when your relatives are under occupation and you know what was done to the unarmed civilians in Bucha, things look different—just a little bit.

It was spring. A strong wind was picking up outside. I was driving in a car to my duty station and listening to Bob Dylan:

And how many years can some people exist
Before they're allowed to be free?
Yes, and how many times can a man turn his head
And pretend that he just doesn't see?

And how many ears must one man have
Before he can hear people cry?
Yes, and how many deaths will it take 'til he knows
That too many people have died?

The answer, my friend, is blowin' in the wind
The answer is blowin' in the wind.

Either you're fighting and risking getting killed by missile shards, or you aren't fighting and are risking getting killed all the same: with a sack over your head, with your hands tied behind your back. At any moment and for any reason, you can end up being tortured.

Fatigue or no fatigue, this is about survival.

¤   ¤   ¤

Nine months into the war, I had the chance to visit my children abroad. This was a very exceptional case for an enlisted soldier. I had been invited to a literary festival as a presenting author. My approval for international travel came all the way from the Commander-in-Chief of the Armed Forces of Ukraine: I was issued a letter signed by General Zaluzhnyi.

Whether or not I would be allowed to attend was decided at the last moment, and in the leadup, I was tormented by what felt like a rupture between two worlds. At lunchtime, you're still patrolling a military facility, unbathed and with a weapon, and two hours later you're speeding in a car to hand in your Kalashnikov rifle, and then the commander personally drives you just as you are, in a dirty uniform, to a bus. You arrive twenty minutes before the plane is scheduled to depart. In Kraków, at the airport closest to the Ukrainian border, everyone is clean. Meanwhile, you're covered in dried mud. Within less than twenty-four hours, you're walking around a European city adorned with Christmas lighting.

Those few days that I was given with them, I dragged my children along with me to all of my literary events, so as to not miss a single hour with them. In the evenings, I didn't step away from them—so as to hug, caress, and kiss them as much as possible. I slept in the same bed as them. I was simultaneously happy that I was with them right then and miserable about how much I had lost because of Gloom's invasion.

Life as it was before the twenty-fourth had ceased to exist. It wouldn't resume if I broke every applicable law, ran away, and stayed illegally in the European Union—which, of course, I never considered.

Riding around a Western European city decorated for Christmas on a tram, I kept feeling that everything I was experiencing there was somehow flat and superficial—not completely real. What I was watching beyond the glass of the tram window was just adornment: decoration fashioned out of shiny foil and colored plastic.

If in the first weeks, as a result of shock, the Russian invasion of Ukraine seemed unreal to me, now peace seemed fictitious and fragile. The residents of peaceful countries didn't appreciate this fragility. I found myself wondering how that guy in Paris who looked like me was doing.

I returned to Ukraine a week later. I walked back to my military unit through the small town where I was serving at the time. It was a cold evening in late November. The town was almost completely dark because of Russia's bombing of our power stations. Only here and there, diesel generators hummed and individual windows shone dimly. I walked in the darkness and sensed that, for me, this was the only reality. This was true existence, the entire depth of being.

Yet I continue to be acutely aware that this is a depth of being a psychologically healthy person would never have chosen voluntarily.